Developments in Geotechnical Engineering 11

DAMS AND EARTHQUAKES

Further titles in this series:

Developments in Geotechnical Engineering 11

DAMS AND EARTHQUAKES

by

HARSH K. GUPTA

Scientist,
National Geophysical Research Institute,
Hyderabad, India

and

Institute for Geosciences,
University of Texas at Dallas, U.S.A.

and

B.K. RASTOGI

Scientist,
National Geophysical Research Institute,
Hyderabad, India

ELSEVIER SCIENTIFIC PUBLISHING COMPANY
AMSTERDAM - OXFORD - NEW YORK   1976

ELSEVIER SCIENTIFIC PUBLISHING COMPANY
335 Jan van Galenstraat
P.O. Box 211, Amsterdam, The Netherlands

AMERICAN ELSEVIER PUBLISHING COMPANY, INC.
52 Vanderbilt Avenue
New York, New York 10017

ISBN 0-444-41330-8

Printed in The Netherlands

To Our Teachers
J. Singh and B.P. Saha

AVANT PROPOS

Parmi les problèmes qui se posent au géophysicien devant la création de plus en plus fréquente des lacs artificiels retenus derrière les grands barrages les phénomènes de séismicité provoqués par le remplissage de certains de ces réservoirs ont particulièrement retenu l'attention. Il n'est peut être pas inutile de faire un rapide historique du développement des recherches dans ce domaine.

Le 10 décembre 1967, un séisme destructeur se produisait en Inde dans une région considérée jusqu'alors comme aséismique; ce séisme entraînait de lourdes pertes de vies humaines et causait des dégâts matériels considérables; l'épicentre, confirmé par les calculs faits au Bureau central international de Séismologie, coïncidait presque exactement avec le barrage de Koyna qui fut sérieusement endommagé par la secousse. L'exemple alors récent du séisme survenu en 1966 à proximité du barrage de Kremasta en Grèce, celui beaucoup plus ancien de l'activité séismique qui avait accompagné le remplissage du lac Mead formé sur le Colorado par le barrage Hoover, activité séismique si bien décrite par Carder dès 1945, des observations analogues que nous avions pu faire en France en 1963 m'incitaient immédiatement à voir une relation de cause à effet entre le remplissage du lac de Koynanagar et le séisme du 10 décembre 1967.

A la suite de l'interview que j'accordais quelques jours plus tard à la journaliste scientifique du *Monde* (Paris), ce journal publiait le 29 décembre 1967 un article intitulé "Les tremblements de terre peuvent avoir pour origine les travaux de l'homme", article dans lequel se trouvaient résumés les renseignements alors en ma possession concernant l'activité séismique au voisinage de certains grands lacs-barrages. Dès le lendemain, à la lecture de cet article, le rédacteur en chef du *New Scientist* me demandait par télégramme un mémoire plus détaillé sur ce sujet. Ce mémoire, traduit en anglais, paraissait le 11 juillet 1968 sous le titre "Fill a lake, start an earthquake", titre sans nuance qui dépassait de beaucoup la pensée que j'exprimais dans le texte.

En janvier 1969, devant la quatrième conférence mondiale de génie paraséismique réunie à Santiago-du-Chili, j'essayais d'établir une comparaison détaillée entre les différents cas observés; à cette époque bien des chercheurs géologues ou géophysiciens restaient encore sceptiques quant à la possibilité de séismes engendrés par le remplissage de lacs-barrages, sans doute parce que les quelques cas connus avaient été étudiés isolément et qu'une coïncidence fortuite entre secousses et remplissage pouvait être retenue raisonnablement.

C'était en particulier la conclusion formulée dans plusieurs des articles ou rapports publiés après le séisme de Koyna. Un comité d'experts n'écrivait-il pas en 1968: "C'est l'opinion du Comité que le réservoir n'est pas responsable des deux grands séismes de septembre et décembre 1967."

Cette coïncidence qui cependant se répétait dans chacun des cas étudiés méritait une étude comparative qui ne laissait bientôt plus de doute sur la réalité du phénomène.

Le problème posé parut suffisamment intéressant tant au point de vue théorique qu'au point de vue pratique pour que l'UNESCO organise en 1970 un "groupe de travail sur les phénomènes sismiques associés à la mise en eau de grandes retenues".

Des travaux de plus en plus nombreux et de plus en plus détaillés, stimulés par les observations faites à Denver d'une activité séismique provoquée par l'injection de fluide sous pression dans un puits profond, et encouragés aussi par d'autres observations analogues faites dans les champs pétrolifères de Rangely (Colorado) et de Lacq (France), ont été publiées; la liste bibliographique s'allonge d'année en année.

Harsh K. Gupta et B.K. Rastogi avaient eu l'occasion d'étudier sur place les effets du séisme de Koyna et d'en entreprendre l'examen détaillé des enregistrements. Ces deux auteurs étaient donc particulièrement qualifiés pour rédiger un ouvrage rassemblant et commentant de façon approfondie les observations faites en différents points du monde à l'occasion du remplissage de certains lacs-réservoirs.

J'attire spécialement l'attention sur le chapitre de l'ouvrage qui est consacré au rôle joué par les pressions dans les couches poreuses. C'est un problème important et nouveau dont on saisit chaque jour davantage les conséquences pour mieux comprendre le mécanisme des séismes et même pour s'engager dans la voie de la lutte directe contre les séismes naturels.

Les ingénieurs qui se préoccupent d'assurer la sécurité des grands barrages trouveront dans le dernier chapitre des recommandations utiles et des indications précises sur les méthodes de contrôle de l'activité séismique avant, pendant et après le remplissage d'un lac-réservoir, en particulier par l'installation d'un réseau de stations séismographiques mobiles.

L'ouvrage, accompagné d'une bibliographie abondante, met ainsi à la disposition des chercheurs une documentation précieuse et considérable. Il faut en remercier les auteurs.

31 décembre 1974

J.P. ROTHÉ
Directeur du Bureau central international
de Séismologie
Strasbourg

FOREWORD

Among the problems facing the geophysicist as a result of the more and more frequent creation of artificial lakes formed by large dams, the phenomena of seismicity provoked by the filling of these reservoirs have received particular attention. It is perhaps useful to undertake a brief historical review of the research developments in this domain.

On December 10, 1967, a destructive earthquake occurred in India, in a region considered until then as being aseismic; this earthquake brought with it a heavy loss of human life and caused considerable material damage; the epicenter, as confirmed by calculations made at the "Bureau Central International de Séismologie" in Strasbourg, coincided almost exactly with the Koyna Dam, which was seriously damaged by this shock. The example of the earthquake which occurred in 1966 in the proximity of the Kremasta Dam in Greece, the much older case of seismic activity which accompanied the filling of Lake Mead formed on the Colorado River by Hoover Dam, the seismic activity so well described since 1945 by Carder, and some analogous observations that we made in France in 1963, immediately led me to see a cause-and-effect relationship between the filling of Lake Koynanagar and the earthquake of December 10, 1967.

On December 29, 1967, the Paris newspaper *Le Monde* published an article, based on an interview that I had given to their scientific correspondent; this was entitled "Les tremblements de terre peuvent avoir pour origine les travaux de l'homme". This article summarized the accounts in my possession of seismic activity in the vicinity of certain large reservoir dams. The next day, upon reading this article, the editor in chief of *New Scientist* asked me by telegram for a more detailed report on this subject. This report, translated into English, appeared on July 11, 1968, under the title "Fill a lake, start an earthquake", a title lacking nuance that overstepped the thoughts that I had expressed in the article.

In January 1969, at the Fourth World Conference on Earthquake Engineering in Santiago, Chile, I attempted to establish a detailed comparison of the different cases that had been observed. At that time a good many geologists and geophysicists were still skeptical about the possibility of earthquakes being engendered by the filling of reservoirs formed by dams, undoubtedly because the several known cases had been studied individually and therefore a fortuitous coincidence between the shocks and the filling could be considered reasonable. The latter reason, in particular, was the conclusion formulated in several of the reports published after the Koyna earthquake.

In fact, a Committee of Experts wrote in 1968: "It is the view of the Committee, therefore, that the reservoir was not responsible for the major shocks of September and December 1967".

This coincidence, which was subsequently repeated when each case was studied separately, merited a more complete comparative study which today leaves no doubt about the reality of this phenomenon.

This problém seemed interesting enough theoretically and practically for UNESCO to organise in 1970 a "Working Group on Seismic Phenomena Associated with Large Reservoirs".

More and more numerous studies in ever greater detail have been published, stimulated by the observations made near Denver of seismic activity provoked by the injection of fluids under pressure into a deep well and also stimulated by other analogous observations made in the petroleum fields of Rangely (Colorado) and Lacq (France); the bibliographical list grows from year to year.

H. Gupta and B.K. Rastogi had the opportunity to study the effects of the Koyna earthquake on the spot and to undertake the detailed examination of seismographic records. These two authors were particularly qualified to prepare a work assembling and commenting on, in a thorough manner, the observations made in different parts in the world on the occasion of the filling of certain large reservoirs.

I draw special attention to the chapter in the book devoted to the part played by pore pressure in inducing earthquakes. It is an important and new problem whose consequences will lead to a better understanding of the mechanism of earthquakes and even to the opening of the way to a direct fight against natural earthquakes.

Engineers responsible for assuring the security of large dams will find in the last chapter useful recommendations and precise indications for the methods of recording seismic activity, before, during, and after the filling of reservoir dams, in particular by the installation of a network of portable seismographs.

The book, together with its abundant bibliography, puts at the disposal of scientists a precious and an extensive documentation.

For this the authors must be thanked.

31 December 1974

J.P. ROTHÉ
Directeur du Bureau Central International
de Séismologie
Strasbourg

PREFACE

There are about thirty cases where the initiation or enhancement of seismic activity has been well evidenced following the impounding of reservoirs behind large dams. Earthquakes at some of these reservoirs have been destructive, exceeding magnitude 6. The one at Koyna in India on December 10, 1967, levelled the Koyna Project Township, claimed over 200 human lives, injured about 1,500 and rendered thousands homeless. These damaging reservoir-associated earthquakes drew the attention of engineers and earth scientists all over the world, giving impetus to the study of the phenomenon of reservoir-associated earthquakes.

This book aims at summing up the present-day knowledge on earthquakes associated with large artificial lakes. A general introduction to the subject is given in the first chapter. The assessment of the focal parameters and macroseismic effects of the Koyna earthquake is presented in the second chapter. Instrumental and macroseismic data support the thesis that the Koyna earthquake was a multiple seismic event. The third chapter deals with the geology, hydrology and seismicity of all the known seismic reservoir sites. Three known cases of induced seismicity following the fluid injection in deep wells are also similarly treated. A possible correlation between the reservoir level/ volume of the injected fluid, and the tremor frequency has also been described in this chapter. The characteristic seismic features of reservoir-associated earthquakes and how they reflect upon the changes in mechanical properties of rock masses near the reservoirs are described in the fourth chapter. Procedural details for the calculation of the stresses added by a reservoir are outlined in the fifth chapter. Theoretical and laboratory experiments on the effect of pore pressure in causing shear failure form the subject of the sixth chapter. The part played by increased pore-fluid pressures in triggering the earthquakes at Denver, Rangely, Kariba, Kremasta and Koyna has been included. Recent developments in fluid-flow stress analysis and in-situ measurements of stresses also have been described briefly in this chapter. It has not been possible to cover all the topics on pore-fluid pressure in detail. Only the salient features which are useful in understanding the reservoir-associated earthquakes have been described. Dam-site investigations and suggestions for the seismic surveillance of the reservoir area are included in Chapter 7. Why certain large reservoirs are aseismic is also commented upon in this chapter.

As this is the first book on the subject, efforts have been made to cover most of the related topics and to present the different view points.

We are extremely grateful to many investigators for sending us their pre-prints/unpublished reports which have been useful. Credit for illustrative material is mostly given in the legend of the figures, but many institutions and individuals have been so helpful that we wish to acknowledge their kindness here. Among the institutions are: Irrigation and Power Department, Government of Maharashtra, India; Butterworth & Co. (Publishers) Ltd.; Geological Survey of India; *Nature*; *Science*; *New Scientist*; Seismological Society of America; Geological Society of America; American Geophysical Union; American Association of Petroleum Geologists; Bureau of Reclamation, U.S. Department of the Interior; Geological Survey of the Republic of South Africa; Geological Society of South Africa and Royal Astronomical Society. Among the individuals are: R.D. Adams, R.E. Anderson, G. Bond, V.R. Deuskar, A.G. Galanopoulos, I.V. Gorbunova, D.I. Gough, S.K. Guha, T. Hagiwara, J. Handin, Jai Krishna, L.N. Kailasam, K. Mogi, B.C. Papazachos, C.B. Raleigh, A. Rogers, J.P. Rothé, Y. Sato, D.W. Simpson, D.T. Snow, and H.I.S. Thirlaway.

Dr. Hari Narain has been a constant source of encouragement. We are grateful for his continued interest, useful suggestions, and support of these investigations. Professor J.P. Rothé encouraged us throughout our investigations and offered many useful suggestions to improve the text of the book. Professor Jim Combs deserves our special thanks for his critical comments on the manuscript and his help during the final stages of the production of this book. Further gratitude goes to our colleague Mr. A.N. Nath for his help in preparing the text of this book. Beneficial comments and suggestions from Professor Mark Landisman, Dr. E.M. Fournier d'Albe, Mr. Indra Mohan, Professor Lynn R. Sykes and Mr. T.N. Gowd are acknowledged. We also wish to thank our colleagues from the National Geophysical Research Institute, Hyderabad, and the Institute for Geosciences, the University of Texas at Dallas for their help and advice. The Photography and the Drawing Section of the National Geophysical Research Institute provided assistance in preparing the figures. The able secretarial assistance of Mr. K. Suryaprakasam, Mrs. Charlotte Scott, Mrs. Alice Somerville and Mrs. Jean Davidson is acknowledged.

HARSH K. GUPTA
and
August 1975   B.K. RASTOGI

CONTENTS

Until recently it was thought that only small earthquakes could be associated with artificial lakes. These earthquakes were explained as due to the sagging of the reservoir basement caused by the load of the water and consequent crustal adjustments. Until the early sixties, although an increase in seismicity was noticed at a number of artificial lakes, it did not cause much anxiety in the absence of any damaging earthquakes — the largest earthquakes were of the order of magnitude 5, which occurred at Lake Mead, formed by the Hoover Dam on the Colorado River in the United States of America. During the 1960s, damaging earthquakes occurred near large reservoirs at Kariba in the Zambia—Rhodesia border region, at Kremasta in Greece and at Koyna in India. These earthquakes of magnitude $\geqslant 6$ claimed many human lives and caused much damage locally, drawing worldwide attention. On request from the local governments, UNESCO sent study missions to investigate the earthquakes at Koyna and at Mangla in Pakistan. Realizing the great socio-economic importance of the phenomenon, UNESCO formed a Working Group on "Seismic Phenomena Associated with Large Reservoirs" in 1970. A number of symposia have been organized and some are being planned on this subject by international organizations.

During the first meeting of the UNESCO Working Group in December 1970, reference was made to 30 large reservoirs. In approximately half of these, impounding had been accompanied and followed by seismic activity. In these cases the frequency and intensity of earthquakes were higher than normal for the region, and their foci appeared to be located in the immediate vicinity of the reservoirs. Instances of reservoir-associated seismicity are now well documented (Rothé, 1968, 1969, 1970; National Academy of Sciences, U.S.A., 1972; Gupta et al., 1972a, 1973). Many more such examples are now being discovered.

The cause of these earthquakes has been given by several workers. Westergaard and Adkins (1934) advanced the hypothesis that sagging of the reservoir basin due to water loading and the consequent readjustments of the underlying substratum are responsible for the geotectonic activity in the reservoir area. Carder (1945) for the first time pointed out that the water load of Lake Mead in the United States has reactivated the pre-existing faults in the area. Depression of the reservoir areas has been noticed at many places. The calculated depression agrees with that observed, as for example in the case of Kariba (Gough and Gough, 1970a). Gough and Gough (1970b) have attributed that the added stresses due to reservoir loading triggers the critically stressed pre-existing faults. Hubbert and Rubey (1959) have drawn attention to the part played by fluid pressure in overthrust faulting. In the last few years, the importance of the increase of pore pressure following reservoir impoundment has been highlighted.

Earthquakes are caused by shear fracturing of rocks. The shear strength of rocks is related to the ratio of the shear stress along the fault to the normal effective stress across the fault plane. The normal effective stress is equal to the normal stress minus the pore pressure. When the pore pressure increases, the shear stress does not alter, but the effective stress decreases by the same amount. Therefore, the ratio of shear to normal stress increases. If rocks are under an initial shear stress, as is generally true, an increase in fluid pressure can trigger shear failure and cause earthquakes. This theory of effective stress has been tested experimentally and demonstrated in the case of fluid injection under pressure into deep wells at Denver (Evans, 1966) and Rangely (Raleigh, 1972) in Colorado, and at Dale (Sykes et al., 1973) in New York. The pore fluids play an important role in the dilatancy phenomenon of rocks, which satisfactorily explains the physical processes taking place prior to most crustal earthquakes (Scholz et al., 1973).

Since the stresses caused by impounded water are small in comparison with the stresses released in the earthquakes, it must be assumed that the rock masses in question were close to failure before the reservoirs were filled. The added stresses due to the water load seldom exceed 10 bars. For Lake Kariba, which is the largest impounded lake in the world (volume of water $175,000 \times 10^6$ m^3), the maximum vertical normal stress added by the water load has been calculated to be 6.68 bars and the maximum added shear stress has been found to be 2.12 bars. The increase in pore pressures caused by artificial lakes could be a few tens of bars. Handin and Nelson (1973) have estimated an increase in pore pressure of at least 15 bars consequent to the filling of Lake Powell (volume $120,000 \times 10^6$ m^3) in the United States. These stresses are very small compared to the strength of the crystalline rocks, which may be of the order of 1,000 bars but they may be sufficient to cause failure of critically stressed faults. In the Rangely oil field of Colorado, the in-situ measurements of stresses by hydraulic fracturing have been made by Haimson (1972). The three principal stresses obtained were 590 bars (horizontal), 427 bars (vertical, assuming the lithostatic pressure to be 0.23 bar/m depth) and 314 bars (horizontal). From these stresses the normal and shear stress across the known fault along which the earthquakes occurred were calculated by Raleigh et al. (1972) to be 347 bars and 77 bars, respectively. With the help of laboratory measurements on the strength of the rocks in question they estimated that a pore pressure of 257 bars is required for slip to take place. This is very close to the bottom-hole pressure of 275 bars, observed in experimental wells when the earthquakes were more frequent. The seismicity ceased after the pressure in these wells dropped by 35 bars. These experiments demonstrated that seismogenic slip on a metastable fault can be triggered by increasing the pore pressure by only a few tens of bars.

In the case of the Rangely earthquakes, it has been possible to estimate the fluid pressure throughout the earthquake zone, since the bottom-hole pressures were known in a number of wells. At the reservoir sites, it has not been possible to make similar estimates. Moreover, the in-situ stresses and the orientation and strength of the existing faults are also not generally known.

Analytical tools are being developed for the calculation of the inter-dependent parameters, permeability and stress, for an assumed fracture geometry in a rock mass (Morgenstern and Guther, 1972; Noorished et al., 1972; Rodatz and Wittke, 1972). Assuming a fracture geometry in the Rangely oil field, Dieterich et al. (1972) have tried to predict future earth-quakes if fluid injection is resumed.

At present most of the researchers seem to accept Hubbert and Rubey's theory of rock failure in which an increase in pore pressure decreases the strength of the rocks. Other factors which may substantially affect the stress field include thermal stresses due to cool water entering warm rock, and the effect of pressure gradients. Research is in progress to evaluate these effects.

Goguel (1973) has argued that the loading effect of a reservoir, if signifi-cant, should tend to decrease any differential stresses at depth caused by the lithostatic pressure of the adjacent mountain masses, and as such the effect of a reservoir should be one of increasing the stability of the region. Snow (1972) has theoretically shown that in a thrust-fault environment, the filling of a reservoir drives the Mohr circle away from the failure envelope, thereby introducing stability. But so far no definite evidence in this regard is avail-able. Mickey (1973a) reported decreases in seismicity following reservoir load-ing for a distance range of 0—40 km from the two dams on the Colorado River, the Glen Canyon Dam in Arizona and the Flaming Gorge Dam in Utah, which are respectively about 300 and 800 km upstream of the Hoover Dam. But Mickey mentions that the majority of the epicenters lie at dis-tances of 200—350 km from the reservoirs within a tectonic zone containing major NNE-trending faults. Since the impoundment of the reservoirs is un-likely to affect the seismic status at such remote distances, natural forces must be responsible for the changes, if any.

The level of seismicity induced by the reservoirs seems to be affected by several factors. The relevance of each of these factors will vary from case to case. Associated seismic activity is particularly clear when the water in the reservoir is deeper than 100 m (Rothé, 1970). The height of the water, and thus the local level of stress, seems to be more important than the total volume of the reservoir in some cases. The rate of increase of water level and the duration for which high levels are retained also seem to affect induced seismicity (Gupta et al., 1972a). Seismicity has also been found to increase following a rapid emptying at the Pieve di Cadore Dam in Italy (Caloi,

1970). Here alternating rises and decreases of the water level correlated with a distinct increase in seismicity.

The reservoir-associated earthquakes are found to show certain common characteristics which differentiate them from normal earthquakes of the regions concerned. These features also indicate a change in mechanical properties of the adjacent rock masses after reservoir impoundment.

However, in most of the reservoirs, including some very large ones, no seismic activity has been observed. It is believed, therefore, that special geological and hydrological conditions are required for the triggering of earthquakes of engineering importance. Geological studies of the seismic-reservoir sites indicate the presence of competent strata which to some extent become less competent and heterogeneous in the presence of water. In many areas hydraulic continuity to deeper layers has been inferred from the presence of permeable rocks and fissures. Faults, which might have been reactivated, have also been observed in several areas. It has been inferred that the impounding of reservoirs has induced earthquakes in areas which were critically stressed. To have a better understanding of the seismic activity following the impounding of reservoirs, it is particularly important to examine why a majority of the large reservoirs have remained aseismic.

Chapter 2

FOCAL PARAMETERS AND THE MACROSEISMIC EFFECTS
OF THE KOYNA EARTHQUAKE OF DECEMBER 10, 1967

Among the earthquakes associated with the impounding of the artificial lakes and the injection of fluid through deep disposal wells, the Koyna earthquake of December 10, 1967, is the most significant, having claimed about 200 lives, injured over 1,500, and rendered thousands homeless. The Koyna Nager township was in a shambles, and more than 80% of the houses were either completely destroyed, or became uninhabitable. The city of Bombay and its suburbs, 230 km away from the epicenter, were rocked. People were driven by panic to the road for safety, and the non-availability of the hydroelectric power from the Koyna Hydroelectric Project paralyzed industry throughout the area.

The Koyna Dam and the Shivaji Sagar Lake are situated in the Peninsular Shield of India, which had been considered free from any significant seismic activity. The seismic zoning map of India, prepared by the Indian Standards Institution in 1962 and later revised in 1966, showed this region to be aseismic. However, after the Koyna earthquake, some earthquakes have been cited in nearby areas in the historical past, suggesting that, although the Peninsular Shield has been geologically stable, it was wrong to rule out the occurrence of such an earthquake entirely.

Soon after the impounding of the Koyna reservoir in 1962, reports of earth tremors near the dam site began to be prevalent. The frequency of these tremors increased considerably from the middle of 1963 onwards. These tremors were invariably accompanied by sounds similar to those of blasting (Mane, 1967). The strongest of these tremors would rattle windows, disturb utensils, etc. To monitor these earthquakes, a close network of four seismological observatories was established, when reports of felt earthquakes began to be prevalent in the region during 1963. The hypocenters were found to cluster near the lake, at a very shallow depth. Before the December 10, 1967 earthquake, five other earthquakes occurred during 1967 which were strong enough to be recorded by many Indian seismological observatories. The September 13, 1967, earthquake was of magnitude 5.5 and it caused minor damage locally.

The meizoseismal area of the Koyna earthquake was investigated by various workers to assess the damage. The focal parameters have also been ascertained by a number of agencies (Narain and Gupta, 1968a, b, c; Tandon and Chaudhury, 1968; Guha et al., 1968; Geological Survey of India, 1968; Committee of Experts, 1968; Gupta et al., 1969, 1971; Jai Krishna et al.,

1970; and others). The Government of India constituted an Expert Committee consisting of engineers, seismologists, and geologists to investigate various aspects of the earthquake. This committee also included members of the UNESCO earthquake study mission. In the following, we briefly discuss the investigations carried out in the Koyna region by us and others relevant to the focal parameters of the December 10, 1967 earthquake and its macroseismic effects.

EPICENTER AND ORIGIN TIME

The epicenter and the origin time of the Koyna earthquake of December 10, 1967, have been independently determined by a number of agencies and organizations. Some of the results of these computations have been summarized in Table I. Dutta (1969) has statistically tested these solutions. According to Indian Standard Time, the earthquake occurred in the early morning of December 11, 1967, at 04:21 and hence this earthquake also has been often referred to as the December 11 Koyna earthquake.

TABLE I

Focal parameters of the Koyna earthquake of December 10, 1967

Agency	Epicenter		Origin time (GMT)	Magnitude		Depth (km)
	lat. N	long. E		m_b	M_s	
Central Water and Power Research Station (C.W.P.R.S.)	17°31.1'	73°43.9'	22:51:17.0	7.0		12
India Meteorological Department (I.M.D.)	17°22.4'	73°44.8'	22:51:19.0	7.5		8
Moscow Academy of Sciences of the U.S.S.R. (M.O.S.)	17°30'	73°48'	22:51:19.0	6.4	6.5	
Bureau Central International de Séismologie (B.C.I.S.)	17°24'	73°36'	22:51:20.0	6.4		
International Seismological Summary (I.S.S.)	17°32.4'	73°50.4'	22:51:23.2	5.9		depth from P excessively negative
United States Coast and Geodetic Survey (U.S.C.G.S.)	17°39.6'	73°55.8'	22:51:24.3	6.0		33 km restricted

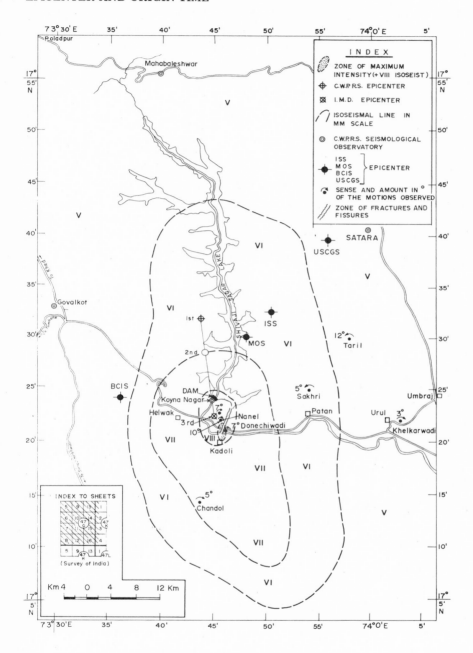

Fig. 1. Map of the Koyna region showing the seismic stations, the epicenters located by various agencies, isoseismals, and the rotational displacements. The later events, located by the analysis of the delay times, are shown as "2nd" and "3rd".

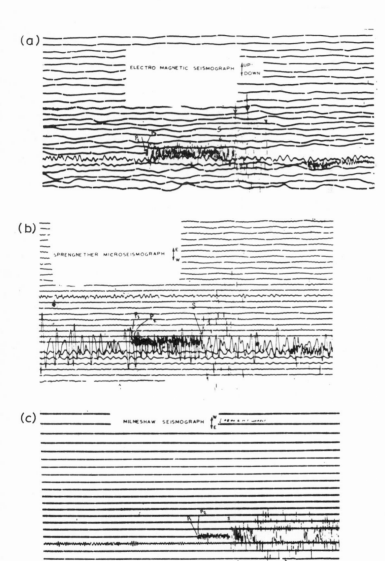

Fig. 2. Sections of the seismograms of the seismic stations at New Delhi (a), Meerut (b) and Port Blair (c) which give support to the view that the Koyna earthquake of December 10, 1967, was a multiple event.

At the Central Water and Power Research Station (C.W.P.R.S.), the origin time and the epicenter have been determined using the data of the four nearby stations at Koyna, Govalkot, Satara and Mahabaleshwar (Fig. 1). All these stations are equipped with sensitive short-period seismographs (T_s = 1 sec, T_g = 1 sec; magnification = 11,000), with recorders operating at a speed of 60 mm/minute, and they are located within a few tens of kilometers of the Koyna Dam. Since the stations are located so close to the focus, they were able to pick up the very first arrivals corresponding to the weak initiation of the earthquake. At the India Meteorological Department (I.M.D.), the focal parameters have been determined using the data of an additional 21 observatories situated at distances varying from 100 to 1,500 km. It is noted that on the seismograms written by the sensitive short-period Benioff seismometers the initial "P"-wave. motion is very weak and is followed by discrete large movements within the next few seconds, indicating the multiplicity of the earthquake. The seismograms shown in Fig. 2 clearly

TABLE II

P-wave residual time (observed minus calculated) for Indian observatories with respect to I.M.D. origin time (magnification and period are given where available)

Station	Obs.—calc. (sec)	Equipment	Magnification	Period (sec)
Poona	—0.6	Benioff	50,000	1.0
Goa	+0.1	Sprengnether	5,000	7.4
Gauribidanur	—0.6	Benioff		1.0
Madras	+0.3	Sprengnether		
Kodaikanal	—0.4	Benioff		1.0
Visakhapatnam	+0.8	Sprengnether	6,000	1.65
Trivandrum	+2.9	Sprengnether	2,500	7.1
Delhi	—0.5	Benioff	50,000	1.0
Rohtak	+0.6	electromagnetic (I.M.D.)		
Sonepat	+0.8	electromagnetic (I.M.D.)		
Bokaro	+1.4	Sprengnether	5,000	7.3
Dehra Dun	+0.5	Sprengnether	1,500	15
Bhakra	0	electromagnetic (H)	5,600	1
Mukerian	0	Hagiwara electromagnetic seismograph		
Pong	+0.7	Hagiwara electromagnetic seismograph		
Jwalamukhi	+0.4	Hagiwara electromagnetic seismograph		
Calcutta	+2.1	Sprengnether	1,000	7
Dalhausi	0	Hagiwara electromagnetic seismograph		
Chatra	0	Benioff		
Shillong	—3.3	Benioff		
Port Blair	—4.7	Benioff		

show multiple "P" phases. The different values of origin time and epicenter obtained for the Koyna earthquake fit in well with a multiple-foci model as discussed in the following. The origin time determined by the I.M.D. is 2 seconds later than the C.W.P.R.S. origin time. Apparently, the distances of these observatories from the epicenter being too large, the first weak initial motion was not picked up everywhere. Table II shows the residuals (observed minus calculated) for the Indian observatories with respect to the I.M.D. origin time. Observatories equipped with sensitive short-period seismographs show negative residuals; those equipped with less sensitive instruments show positive residuals. The total positive residual is 10.6 seconds and the negative 10.1 seconds. This indicates the occurrence of a stronger second event about 3—4 seconds after the initiation, which has been picked up as the first arrival by the observatories operating less sensitive seismographs.

A still later origin time has been reported by the United States Coast and Geodetic Survey (U.S.C.G.S.) and the International Seismological Summary (I.S.S.), where the data of hundreds of the seismological stations distributed throughout the world have been used. Obviously, the arrivals corresponding to the initiation were too weak to be recorded at distant observatories, and the least-squares solution gave a later origin time. Tandon and Chaudhury (1968) have given the residual (observed minus calculated) travel times with respect to the I.M.D. origin time for 188 world stations. Most of them have positive residuals and the total positive residual adds up to 300.9 seconds, while the total negative residual is only 34.1 seconds. Similar residuals have been given by Tandon (1954) for the Assam earthquake of August 15, 1950 (epicenter: 28.5° N, 96.7° E), where the total positive residual adds up to 51.1 seconds and the negative residual total is 34.1 seconds for a total of 61 stations distributed throughout the world. The paths traversed by the seismic waves for most of the distant observatories did not differ for the two epicenters (the Assam earthquake of August 15, 1950, and the Koyna earthquake of December 10, 1967). Since the same Jeffreys-Bullen travel-time tables have been used, the much larger positive residuals, compared to the negative residuals for the Koyna earthquake, support the idea that the first arrivals for the Koyna earthquake have been missed at most of these observatories.

To verify the inferences drawn regarding the multiplicity of the Koyna earthquake on the basis of P-wave arrival time analysis as discussed above, a detailed analysis of the seismograms from 48 stations and the correlograms from the arrays of seismometers at Warramunga (WRA), Australia, Yellowknife (YKA), Canada, and Eskdalemuir (EKA), United Kingdom, was later carried out (Gupta et al., 1971). Fig. 3 shows the correlograms of the Koyna earthquake obtained at the three seismic-array stations mentioned above. The difference between the correlograms obtained for simple earthquakes and explosions from those of complex earthquakes has been pointed out by Thirlaway (1963). The former are characterized by a large buildup of energy

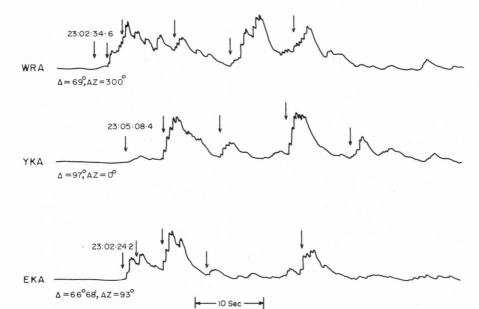

Fig. 3. Correlograms of the seismic arrays at Warramunga (WRA), Yellowknife (YKA) and Eskdalemuir (EKA). Six events are marked by arrows.

which gradually dies down, while the latter have conspicuous bursts of energy at short time intervals (Fig. 4). Following a weak initiation, the beginnings of significant energy buildups, which would correspond to later events, are marked by arrows in Fig. 3; and the time differences with respect to the initiation have been tabulated in Table III. As is evident from this table, the fourth and the fifth events are quite well correlated on these correlograms.

Figs. 5 and 6 show some of the conventional short- and long-period seismograms in which the distinct arrivals following the weak initiations have

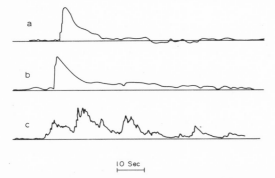

Fig. 4. Correlograms for an explosion (a), a simple earthquake (b), and a complex earthquake (c) (after Thirlaway, 1963).

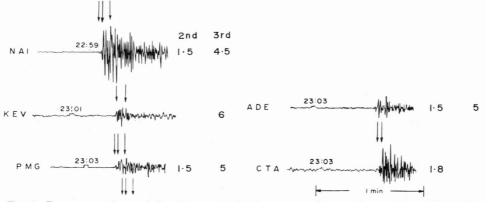

Fig. 5. P-wave portions of the short-period seismograms from a few stations. The initiation and later events are marked by arrows and the delay times (in sec) are noted alongside.

been marked by arrows and their time lag with respect to the initiation has been indicated. The second and the third events, which occurred approximately 2 and 5 seconds later, are most conspicuous and could easily be identified and well correlated on both the seismograms and correlograms. To locate these events with respect to the initiation of the seismic activity at the C.W.P.R.S. epicenter their approximate distance has been calculated using

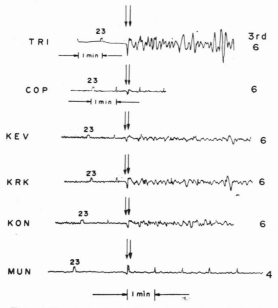

Fig. 6. Tracing of the P-wave portion of the long-period seismograms from some of the stations. The initiation and the third event are marked by arrows and the delay times (in sec) are noted alongside.

TABLE III

Delay time with reference to initiation (event No. 1)

Event No.	Delay time (sec)		
	WRA	YKA	EKA
2	1.8	—	2.0
3	4.0	5.6	5.8
4	11.8	13.9	12.4
5	20.0	23.6	26.2
6	29.4	33.2	—

the following relation (Wyss and Brune, 1967), after arranging the recording stations in azimuth:

$$\Delta i = \frac{t_i}{(1/\beta) + \cos (\theta - \phi) dp/d\Delta}$$

where t_i is the time interval in seconds after the first arrival read from the seismograms; β is the assumed propagation velocity of the rupture; $dp/d\Delta$ is the slope of the travel-time curve at Δ; θ is the azimuth from the initial epicenter to the station; and ϕ is the azimuth from any later event to the initial epicenter, a parameter (effectively the direction in which the rupture propagated).

The factors affecting the time lag which occurred before the later events appeared on the seismograms include the azimuth and the epicentral distance of the recording station in addition to the position and the time difference of the later events with respect to the initiation. Adopting a rupture velocity of 3.5 km/sec, taking $dp/d\Delta$ from the Jeffreys-Bullen travel-time curves, and assuming $\phi = 352°$ which is consistent with fault-plane solutions determined for this earthquake, values of Δi have been computed from the time lags with which the signals from the later events arrive at various stations at different azimuths. Using Gutenberg's sine-curve method, the second event is located at a distance of 6 km, and the third event at a distance of 17 km with respect to the C.W.P.R.S. epicenter. The negligible scatter (Figs. 7 and 8) shows that the value of 352° chosen for ϕ is reasonable. The epicenters of the first, second, and third events are plotted in Fig. 1.

The above analysis supports the thesis that the Koyna earthquake was very likely a multiple seismic event, of which as many as six sub-events could be detected on the seismograms and the correlograms. Among these, the second and the third are most conspicuous on conventional seismograms. The rupture initiated at the C.W.P.R.S. focus and then proceeded towards the south for at least the next 6 seconds with an average velocity of 3.4 km/sec (Table IV).

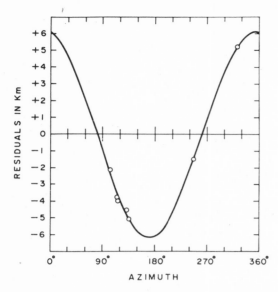

Fig. 7. Gutenberg sine-curve plot for the location of the second event.

The Koyna earthquake has also been found to be a multiple seismic event in another detailed analysis carried out independently by Gorbunova et al. (1970). They have used the first-arrival time data from observatories in the following four epicentral distance ranges to obtain the focal parameters of the Koyna earthquake: (1) only Indian stations located near the epicenter ($\Delta \leqslant 2°$); (2) intermediate stations ($2° < \Delta \leqslant 20°$); (3) stations from all over the world (107 in number) in a distance range of 0.03° to 100°, and (4) distant stations ($\Delta > 20°$).

The focal parameters thus determined differ considerably for the four sets of computations (Table V), and the epicenters do not coincide (Fig. 9).

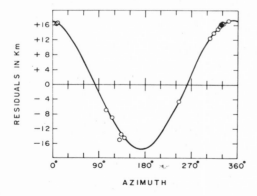

Fig. 8. Gutenberg sine-curve plot for the location of the third event.

TABLE IV

Parameters obtained for second and third events

Event No.	Time lag (sec)	Distance to first event (km)	Velocity* (km/sec)
2	1.8	6	3.6
3	5.4	17	3.2

* Average rupture velocity = 3.4 km/sec.

TABLE V

Focal parameter obtained by Gorbunova et al. (1970) for the Koyna earthquake of December 10, 1967, using different sets of observations

System of stations	Origin time (GMT)	Epicenter		Depth (km)	No. of stations
		lat. N	long. E		
1	22:51:20.0	17°23′	73°45′	0	4
2	22:51:17.8	17°24.6′	73°46.8′	13	13
3	22:51:19.1	17°27.6′	73°48.0′	0	107
4	22:51:28.1	17°33.6′	73°47.4′	64	90

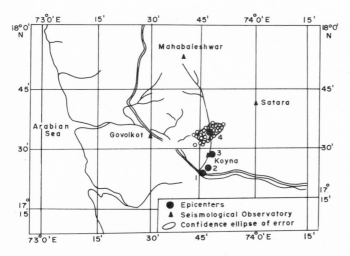

Fig. 9. Epicenters located by Gorbunova et al. (1970) for the four different sets of observations (see text).

Gorbunova et al. (1970) note that the difference in the locations of epicenters 1 and 4 cannot be due to errors in location. They have calculated a confidence ellipse of errors for epicenter 4 (shown in Fig. 9). Twenty sets were made from the data of distant stations, each having at least 40 observations, and the epicenters thus located are found to surround epicenter 4 quite evenly (Fig. 9). The maximum semi-axial value of the confidence ellipse of errors does not exceed 5 km. On the basis of this analysis, they have suggested the horizontal extent of the focal region to be of the order of 20—25 km, which is the distance between their epicenters 1 and 4.

FOCAL DEPTH

As can be noted from Table I, different focal depths have been assigned to the Koyna earthquake by different agencies. Tandon and Chaudhury (1968) have estimated the depth between 9 km and 32 km using empirical relations for different sets of magnitude and maximum intensity values. Considering that heavy damage has been restricted to a small area while the area where the shock was felt was comparatively much larger, they inferred either the occurrence of two earthquakes — one at a shallow depth of about 8—10 km which caused the local damage, and the other at about 30 km which was responsible for the huge area where the earthquake was felt — or the extension of the propagating fracture from about 12—20 km depth to about 35 km depth. Gorbunova et al. (1970), by minimizing the station residual values, have obtained focal depths of 0 and 64 km for hypocenters 1 and 4 (Ta-

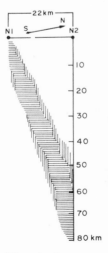

Fig. 10. Source propagation of the Koyna earthquake of December 10, 1967 (after Gorbunova et al., 1970).

ble V). They have then calculated $\delta i = |fi'/fi''|$, where fi' is the residual of the ith station from hypocenter 4, and fi'' is the residual of the same station from the chosen hypocenter 1. They argued that if all the stations of a system registered first arrivals of seismic waves radiated from the same point of a focal region, i.e. hypocenter 1, then all the values of the station residuals would be smaller than those for hypocenter 4 and δi would be >1. If $\delta i \leqslant 1$, it may be understood that there exists another point of radiation within the focal region giving a minimal residual value. Under these circumstances, there would be a discrepancy between the values of δi for epicenter 1 in the data of the stations located at different epicentral distances. After detailed analysis, Gorbunova et al. (1970) concluded that the nature of the motion in the focal region was characterized by a slip along a nearly vertical fault plane at a velocity of about 8 km/sec. The apparent horizontal velocity of the fault has been approximated to be about 3 km/sec. This is consistent with the rupture velocity of 3.4 km determined from analysis of the time interval between the weak initiation and stronger later arrivals as discussed earlier. Fig. 10 gives the scheme of the origin of the Koyna earthquake as deduced by these authors.

MAGNITUDE

As can be seen from Table I, with the exception of the I.M.D. and the C.W.P.R.S., who have given the extremely high body-wave magnitudes of 7.5 and 7.0 respectively for this earthquake from Wood-Anderson records, the values given by the other agencies vary from 6 to 6.5. Narain and Gupta (1968a, b), on the basis of the study of macroseismic effects of the earthquake, estimated a shallow focus and a rather lower magnitude than that given by the I.M.D. and the C.W.P.R.S. They obtained a surface-wave magnitude value of 6.2 from the analysis of seismograms recorded at a number of observatories. The observatory at Uppsala reported a magnitude of 6.25 for this earthquake. Trifunac and Brune (1970) have pointed out that for multiple-event earthquakes, the surface-wave magnitude calculated from waves of 20-second periods tends to be larger than the body-wave magnitude. This is due to the fact that the relationship of magnitude versus moment is not single-valued. An increase in the number of the closely spaced events tends to increase the total moment but not necessarily the maximum amplitude. The most representative values of body- and the surface-wave magnitudes for the Koyna earthquake are 6.0 and 6.3 respectively. From the relation $M = 1.59m - 3.97$ (Richter, 1958), where M and m are the surface- and the body-wave magnitudes respectively, the surface-wave magnitude of the Koyna earthquake should have been 5.6 instead of 6.3. The relatively higher surface-wave magnitude for the Koyna earthquake supports its multiple nature.

INFERENCES DRAWN FROM FIELD EVIDENCE AND STRONG-MOTION DATA

The I.M.D. epicenter lies very close to the zone of maximum intensity of VIII on the Modified Mercalli intensity scale (Fig. 1). A very significant and interesting observation has been made in terms of the recorded rotational displacements. Seven such movements have been very well documented and reported by the Geological Survey of India (1968). Fig. 11 shows one such typical rotational movement which was photographed and recorded by us. The sense and the amount of these rotational movements are given in Table VI and plotted in Fig. 1. The sense of movement on the western side of the line joining the C.W.P.R.S. epicenters and the zone of maximum intensity are all clockwise, whereas they are anticlockwise on the eastern side of this line with the exception of the observation at Khelkarwadi. This pattern of rotational movements can be expected if we consider the C.W.P.R.S. focus as the initiating point from which rupture continued in an approximately southern direction up to the I.M.D. epicenter and probably beyond.

The accelerograms for the Koyna earthquake from the strong-motion seismographs located in a gallery at about the mid-height of the dam are reproduced in Fig. 12a, which has been adopted from Jai Krishna et al. (1969). The horizontal component strong-motion seismographs are oriented in directions parallel and perpendicular to the dam. Fig. 12b shows the particle

Fig. 11. Rotational displacement of a pillar at Donchiwadi (photograph by H.K. Gupta).

TABLE VI

List of the locations where rotational movements have been measured with their magnitudes and directions

Location	Rotation		Recorded by
Sakhri	$5°$	anticlockwise	
Tarli	$12°$	anticlockwise	
Wai	$14°$	anticlockwise	
Nanel	$7°$	clockwise	Geological Survey of India
Dondchiwadi (N)	$10°$	clockwise	
Chandol	$5°$	clockwise	
Khelkarwadi	$3°$	clockwise	
Dondchiwadi (S)	$7°$	anticlockwise	National Geophysical Research Institute

motion plotted for about 11 seconds, revealing a predominantly N—S motion. This observed trend of particle motion is consistent with the disturbance produced by longitudinal waves from a source moving from north to south or vice versa, and recorded at the Koyna Dam by horizontal component accelerographs with their existing orientations.

Investigations of the macroseismic effects of the Koyna earthquake have revealed a very localized area suffering damage corresponding to VIII on the Modified Mercalli (MM) intensity scale. However, the earthquake was felt over distances as large as 700 km. Tandon and Chaudhury (1968) and

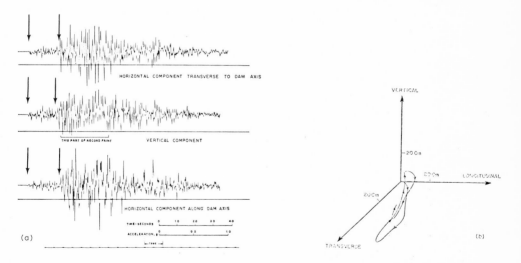

Fig. 12. (a) Accelerograms for the Koyna earthquake recorded at the dam (adopted from Jai Krishna et al., 1969). The initiation and second event are marked by arrows. (b) Particle motion, obtained from the accelerograms, plotted for about 11 initial seconds.

Gorbunova et al. (1970) have interpreted this as support for their focal
models of the Koyna earthquake, discussed earlier. According to them, the
localized damage in the Koyna township was caused by the shallow-focus
event, and the large felt area was due to the later, deeper event. However, it
has been noted from the study of the three most recent large earthquakes in
Peninsular India, viz. the Koyna earthquake of December 10, 1967; the
Godavari Valley earthquake of April 13, 1969, of magnitude 5.7; and the
Broach earthquake of March 23, 1970, of magnitude 5.4; that the area of
perception has been unusually large for the moderate-magnitude earthquakes
(Gupta et al., 1969, 1970, 1972c). Both, the Broach earthquake and the
Godavari Valley earthquake had undisputed shallow depths. The large areas
of perceptibility are probably attributable to the efficient transmission of
seismic energy across the Peninsular Shield of India.

  From the investigations carried out for the Koyna earthquake as discussed
in the previous sections, it could be summarized that the Koyna earthquake
was a multiple event. The initial rupturing initiated somewhere close to the
C.W.P.R.S. focus, then continued southwards to the I.M.D. hypocenter for
the first 5—6 seconds. Its later continuation, again generally northwards as
found by Gorbunova et al. (1970), for the next 4.5 seconds, with the rupture penetrating to a depth of 60—70 km cannot be ruled out. Guha et al.
(1968) observed that there is a general tendency for the tremors in the
region to begin approximately 20 km upstream, north of the dam; then the

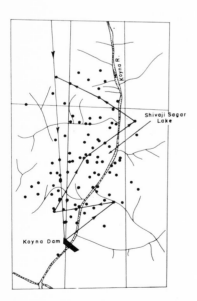

Fig. 13. Epicenters located around the lake. Inferred migration is shown by the arrows
(after Guha et al., 1968).

epicenters shift towards the south, before going back again to the north (Fig. 13). This cycle has been repeated in the past. An event 12—14 seconds after the initiation has also been identified in the correlograms. If the whole cycle were to be repeated in a very short time with events occurring closely together, going from north to south and back again, characteristics similar to the Koyna earthquake could be observed.

ISOSEISMAL MAPS

The Koyna earthquake of December 10, 1967, shook the western half of Peninsular India. It was distinctly felt as far as to Surat and Ujjain in the north, Nagpur and Hyderabad in the east, and Calicut and Bangalore in the south and the southeast. Intensities allocated to an earthquake-affected area, especially in the macroseismic zone which includes different types of construction, are likely to differ subjectively from one investigation to the

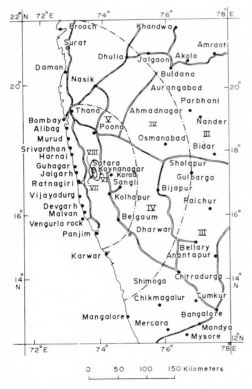

Fig. 14. Map showing the isoseismals in Western India (modified from the Geological Survey of India, 1968).

other. The Koyna region is no exception. The isoseismal maps for this earthquake have been prepared by four different Indian organizations. These maps have similar general features but differ appreciably in their minor details. The isoseismals drawn by the Geological Survey of India (1968) show an area of 50 km^2 included within isoseismal VIII (MM scale), whereas those drawn by the School of Research and Training in Earthquake Engineering (1968), Roorkee University, have an area of 200 km^2 within isoseismal VIII. The Committee of Experts considered the various versions and Fig. 1 shows the map of higher isoseismals which they accepted. Fig. 14 shows the map of lower isoseismals for Western India as prepared by the Geological Survey of India. This version has been generally accepted by all. In their report, the Officers of the Geological Survey of India observe that the characteristics of the area of maximum damage do not appear to be commensurate with the magnitude of 7.5 on the Richter scale assigned by the I.M.D. to this earthquake. The extent and type of damage are suggestive of a lower magnitude. Similar observations have been reported by others after investigating the damage in Koyna Nagar and the surrounding areas (for example, Narain and Gupta, 1968a; Wadia, 1968; and others). The maximum intensity observed does not exceed VIII on the MM scale. An area of about 100 km^2, elongated in the N—S direction, falls within isoseismal VIII. A zone, about 3 km in length and 1 km in width, of maximum intensity could be delineated (Fig. 1) under this isoseismal. The Koyna Dam is situated in the northern part of this area. Kodali and Dondechiwada villages are in the center of the intensity VIII area, where almost all the houses, built of stone and sun-dried bricks, collapsed. Even in these villages, as well as in other areas falling under intensity VIII, the water supply lines, electric, telegraph and telephone poles remained more or less intact. This is consistent with the view that the intensity did not exceed VIII.

An area of approximately 600 km^2, elliptical in shape, falls within isoseismal VII. The area spreads towards the south of the Koyna Valley and decreases in the E—W direction (Fig. 1). The majority of the houses in this area suffered varying degrees of damage, the degree depending upon the constructional details of each house. Isoseismal VI covers an area of about 2,000 km^2, and is also elliptical, with the major axis oriented in the N—S direction (Fig. 1). Only minor cracks developed in the well-constructed houses in this area. The spread of isoseismals V and IV is very wide, and they are incomplete because of the presence of the Arabian Sea on the west (Fig. 14). However, they also retain the general elliptical trend observed for the higher isoseismals.

ISOFORCE MAP

In addition to the isoseismal maps mentioned above, Jai Krishna et al.

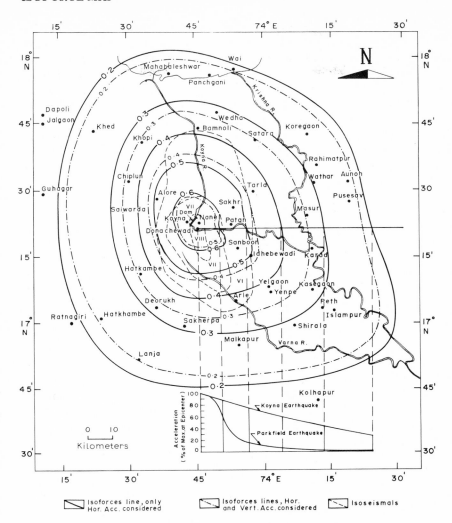

Fig. 15. Isoforce map for the Koyna earthquake (after Jai Krishna et al., 1970).

(1970) have studied the dynamic behavior of many small objects located at various distances from the epicenter, and have prepared an isoforce map for the Koyna region (Fig. 15). They made a quantitative assessment of the forces which acted on engineering structures and many small objects located at various distances from the Koyna earthquake epicenter. In all, about 1,200 observations of sliding, overturning, or the absence of these effects, have been made at several places scattered within 60 km from the Koyna township. Out of these, Jai Krishna et al. (1970) considered 400 observations to be reliable. A systematic analysis of the dynamic behavior has given the lower and upper limits of the peak ground forces at these places. A graphical interpolation between the upper and the lower limits of the ground forces has next been carried out to obtain the distribution of ground acceleration, in the form of attenuation curves in different directions. The points of equal acceleration have been connected to obtain the contours and the map thus obtained has been termed the *isoforce map*. From these studies, Jai Krishna et al. (1970) concluded that the isoforce lines corroborate the seismotectonic features of the area. It is noticed that the attenuation of the ground accelerations is very gradual in the area compared with that reported for other earthquakes. This has been attributed to the efficient transmission of the energy through the peninsular rocks.

FISSURES, CRACKS IN THE GROUND AND ROCK FALLS

The earthquake caused numerous fissures and cracks generally in the soil-covered area and also at some places in highly weathered traps. The most conspicuous was the regular system of fissures and cracks traceable for a distance of about 3 km across the west to east flowing Koyna River between Nanel in the north and Kadoli in the south (Fig. 1). Fissures up to 40 m in length and varying in width from a few centimeters to 40 cm are mostly confined to a 200 m wide zone (trending NNE—SSW). The fissures mostly strike from N10° W to N25° E and are arranged in an en-echelon pattern. Between these fissures, a number of diagonal tensional cracks are observed which trend N10° W to N40° W. Fig. 16a, b shows typical fissure and crack patterns and Fig. 17 shows the fissures which developed in the Karad—Guhagarh road. This place is very close to the crossing of the above-mentioned fissure zone which intersects the Koyna River where it takes a sharp turn to the east. It is necessary to mention that these fissures do not seem to penetrate deeper than 2—3 m. Pits, opened at a number of places, did not show their continuation beyond a depth of 2—3 m.

The diagonal cracks, seen between the ground fissures at many places, are suggestive of shear movement. Star-pattern cracks are also observed, bulging up to 20 cm in height. These bulges vary in shape. At some places they appear as circular heaps while at others they appear as broad flat mounds. Some interesting patterns of en-echelon fissures are observed in loose soil. Fig. 18 shows the edges of the terraced land close to the epicenter falling off.

Rock falls have been observed at many places on several high ridges on either side of the Koyna Valley and the continental divide. These rock falls seem to have been actuated by the relative movement of large blocks of the columnar basalt. The columnar and vertical joints displayed at many places in the flood basalts must have facilitated such happenings since the laterite capping these joints is prone to displacement on shaking. The rock blocks,

Fig. 16. Typical pattern of the cracks developed (a) in the road, and (b) in the ground (photographs by H.K. Gupta).

Fig. 17. Close views of a crack developed in the Karad—Guhagarh road (photographs by H.K. Gupta).

Fig. 18. Damage to the edges of terraced land close to the epicenter (photograph by H.K. Gupta).

while rolling down the steep cliffs and escarpments, brought with them large amounts of the scree material.

DAMAGE TO CIVIL ENGINEERING STRUCTURES

The structures which were damaged by the Koyna earthquake include

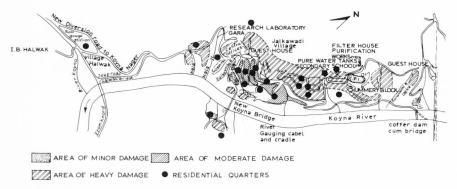

Fig. 19. Damage in the Koyna Nagar township (modified from Committee of Experts, 1968).

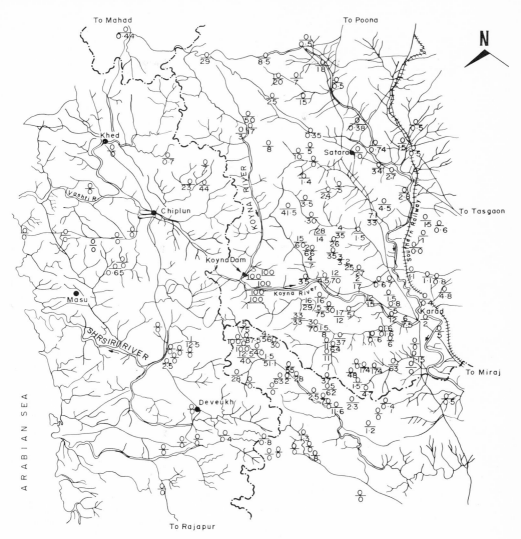

Fig. 20. Numerical representation of the damage in nearby villages. In the ratio a/b, a represents the percentage of totally collapsed houses while b is the percentage of the partially damaged houses (after Committee of Experts, 1968).

buildings, bridges, roads, ports, the Koyna Dam and appurtenant works. Maximum damage has been observed in the Koyna township situated on the western bank of the Koyna River between the dam and Helwak where the river course takes a sharp turn eastwards. Also, almost all the buildings at Donechiwada, Helwak and Nanel villages were destroyed. Fig. 19 shows the damage in the Koyna Nagar township while Fig. 20 gives the percentage of houses which suffered partial or complete damage in the area affected by the Koyna earthquake.

Buildings

On the basis of the constructional details, the houses in the Koyna Nagar township and the surrounding areas could broadly be divided into four categories (Committee of Experts, 1968).

1st category

The houses with thick rubble masonry walls which were constructed using lime or cement suffered the severest damage in spite of having light roofs consisting of timber trusses and corrugated steel or asbestos cement sheeting. The photographs in Fig. 21a show two typical examples of the damage such houses incurred. The heavy damage or complete collapse of this type of structures could be attributed to the heavy weight and low tensile strength, high rigidity, proximity to the epicenter and the lack of good bonding in the rubble masonry walls, especially at the junction of the longitudinal and the cross walls. Buildings with brick walls have suffered less, (Fig. 21b), probably because of the better bonding adopted in their construction.

2nd category

The light and flexible temporary constructions consisting of timber poles and horizontal runners having corrugated iron sheets remained practically unaffected (Fig. 22). This was because these constructions were light and flexible.

3rd category

This category includes the houses consisting of random rubble masonry in mud or laterite blocks and using tiles or asbestos sheets for roofing. The rigid roof framing system rests on the timber posts which are framed into it. A sort of pin-connection effect is provided by the tenons of the timber posts going into mortises in blocks of hard stone. These houses suffered much less damage than the houses belonging to the first category. The damage was mostly confined to the panel walls from which the timber posts occasionally fell. A typical example of failure of the panel walls is documented in Fig. 23.

4th category

Most of the houses in the villages have walls made out of random rubble in mud mortar. As such constructions are inherently extremely vulnerable against lateral movements, these houses suffered heavy damages which were mostly due to the falling or heavy cracking of the walls. Fig. 24 shows such a failure.

Fig. 21. (a) Damage to the houses belonging to the first category (photographs by H.K. Gupta).

Fig. 21. (b) The first-category houses with brick walls suffered less (photographs by H.K. Gupta).

Fig. 22. Houses belonging to the second category remained unaffected (photographs by H.K. Gupta).

Fig. 23. Typical failure of the panel wall in the houses belonging to the third category (photograph by H.K. Gupta).

Fig. 24. Failure of the walls belonging to the fourth-category houses made from random rubble in mud mortar (photograph by H.K. Gupta).

Bridges and roads

The most conspicuous damage inflicted by the Koyna earthquake was the collapse of the 100-year old stone masonry arch bridge on the Helwak—Chiplum road near the Koyna township. The bridge is 5.75 m wide and consists of seven spans. Of these the three central spans, each 18 m in length, collapsed (Fig. 25). The bridge is located within the MM VIII zone, and large earthquake forces in the region are evidenced by the extensive cracking of the soil. Other recently constructed bridges in the same area, such as the Koyna Nagar bridge, (a concrete T-beam bridge having six spans and resting on masonry piers) which lies between the above-mentioned bridge and the Koyna Dam, escaped undamaged. Poor construction and old age seem to be largely responsible for the collapse of the Helwak bridge.

Many culverts on the roads connecting the Koyna township with the nearby towns were damaged (Fig. 26). Severe cracks developed in places on the road. Figs. 16a and 17 show such cracks on the road between Helwak and Patan at a distance of 6 km from Helwak. The level and mile stones were often tilted (Fig. 11) and they sometimes toppled following the sliding of the unmetaled shoulders of the roads.

Fig. 25. The stone masonry arch bridge on the Helwak—Chiplum road collapsed (photograph by H.K. Gupta).

Fig. 26. Damage to the culverts on the roads (photographs by H.K. Gupta).

Ports

As mentioned in the report of the Committee of Experts (1968) on the Koyna earthquakes, some of the nearby ports of the Maharashtra State were slightly damaged by the December 10, 1967 Koyna earthquake. The Jaigarh Port, which is equipped with a class 2 lighthouse, is situated at a distance of 65 km from the Koyna Dam (Fig. 14). The 12-m high lighthouse is located on a hillock about 40 m above sea level. A considerable amount of mercury from the trough, on which the optics of the lighthouse is mounted, spilled following the earthquake and the rotating mechanism was disrupted. Also the mantle of the optics shattered.

Thirty-four posts of the railing of a new cement concrete jetty constructed during 1961 at the Vijayadurg Port, situated at a distance of 100 km from the Koyna Dam, cracked following the Koyna earthquake. Similarly, jetty No. 2 of the Redi Port, which is about 160 km from the Koyna Dam, was damaged.

It is noticeable that a considerable strip of land between the dam and the coast was not affected by the earthquake; however, evidence of damage is present at the above-mentioned three places. It seems that the topography of the pre-trap rocks is mostly responsible for such an uneven damage pattern.

DAMAGE TO THE KOYNA DAM AND APPURTENANT WORKS

Before discussing the damage caused to the Koyna Dam and its appurte-
nants, it is worthwhile to have some idea about the dam's constructional
details. The Koyna Hydroelectric Project is situated in the Maharashtra State
at a distance of 200 km SSE from Bombay. The aim of the project is,
basically, to generate hydroelectric power and provide water for irrigation.
Before construction, the sites of the dam, tunnels, penstocks and the powerhouse were investigated by drilling a large number of boreholes to find the
existence of any significant geological faults. Fig. 27 gives a geological profile
along the tunnel system and the underground powerhouse.

The 85 m tall Koyna Dam is a gravity structure with a length of about
854 m. The dam is constructed of monoliths which are on average 16 m wide
(Fig. 28). The dam is founded on a 30-m thick flow of massive basalt, with
the exception of 200 m of the river section between foot chainages 2575 and
3225 (Fig. 28). A shear or fracture zone running in an almost N—S direction
and cutting the axis of the dam at about 60° was detected. It was encountered in monoliths 12, 13 and 14 and was found to be about 4.5 m wide.
This shear zone consists of broken and slicken-sided rocks and is filled with
gouge. Apparently there was no vertical displacement in the lava flows on
the two sides of the fracture. The shear zone was excavated, filled with
concrete and grouted. An accelerograph has been installed in the gallery
constructed above this treated zone. Foundation treatment for the dam
consisted of shallow grouting to consolidate the foundation rock and deep
grouting to provide a grout curtain on the upstream side, followed by a row
of drainage holes. The dam is constructed of rubble concrete, with the
exception of the last six blocks on the left bank which are of hand-laid
rubble masonry.

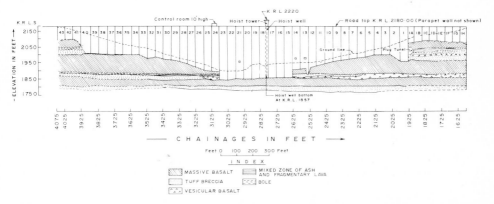

Fig. 27. Foundation geology of the Koyna Dam (modified from Committee of Experts,
1968).

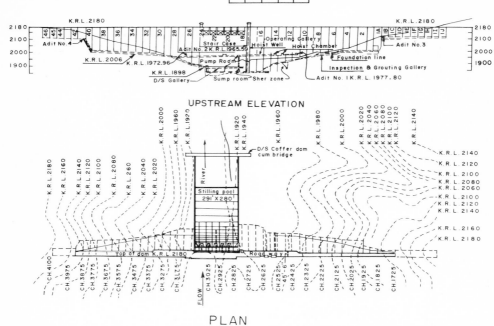

Fig. 28. Plan of the Koyna Dam and the dam's elevation (after Committee of Experts, 1968).

Fig. 29. Damage to the hoist tower located on monolith 18 of the dam (photographs by H.K. Gupta).

Fig. 30. Movement of the unfixed tiles on the span of the Koyna Dam. Note the increase in displacement towards the center of the span from (a) to (c) (photographs by H.K. Gupta).

The 17 m tall and 10 m wide hoist tower of the dam located on monolith 18 suffered a considerable amount of damage. The precast concrete block panelling and the main R.C.C. framework cracked at a number of places (Fig. 29). The unfixed tiles covering the openings on the vertically formed holes between the joint seals across the monoliths on the upstream foot path at the top of the dam were thrown towards the downstream side. The extent of this movement increases as one moves from the ends of the dam to the center (Fig. 30). The compression in the center of the span of the dam is very well documented (Fig. 31).

Tiltmeters and seismometers, installed in a temporary masonry structure on monolith 17, were overturned and dislodged from the table. The structure also suffered some damage at the junction of the rock and the walls. The reinforced concrete walls of the control panel room on monolith 24 cracked horizontally (Fig. 32) along the construction joint of the wall. A steel ladder at the same place was bent in the center (Fig. 33).

The contraction joint between monoliths 26 and 27 seems to have become narrower at the top and wider at the bottom. There is some spalling of the concrete in the parapet at the top near the joints. The spillway bridge, which

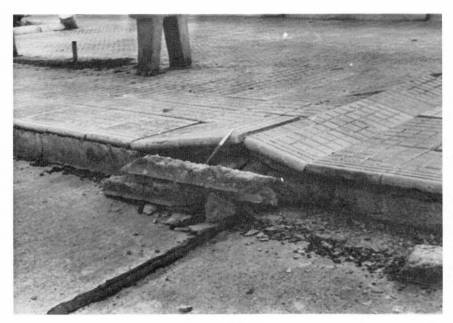

Fig. 31. Compression documented at the center of the Koyna Dam span (photograph by H.K. Gupta).

Fig. 32. Horizontal crack in the concrete wall of the control panel room on monolith 24 (photograph by H.K. Gupta).

Fig. 33. Bent steel ladder at the control panel room situated on monolith 24 (photograph by H.K. Gupta).

has a span of 15 m, suffered a considerable amount of damage. The deck slab cracked longitudinally along the top corner of the downstream T-beam on monoliths 20, 21, 22 and 23. The typical damage is shown in Fig. 34. The piers of the bridge themselves showed some cracking at their junction with the crest of the spillway.

Other important structural damage to the Koyna Dam was the development of horizontal cracks on both the upstream and downstream faces of a number of the monoliths. Fig. 35 shows the approximate locations of these horizontal cracks on the two sides of the dam. Significant leakage on the downstream face of the dam was noticed on a number of monoliths.

Cracks were also noticed in the operation gallery at about mid-height on both the upstream and downstream walls in monoliths 11—19. Similar cracks were also noticed in the foundation gallery. It was also observed that the

Fig. 34. Typical spalling of concrete on the T-beams of the monoliths (photograph by H.K. Gupta).

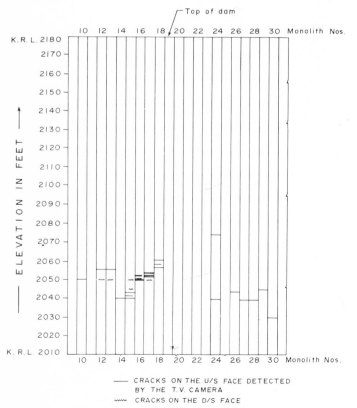

Fig. 35. Approximate location of the horizontal cracks developed on the down and upstream faces of the Koyna Dam (after Committee of Experts, 1968).

amount of water pumped from the sump in the foundation gallery almost doubled after the December 10 earthquake, while the reservoir levels remained the same as before the earthquake.

The above-mentioned description of the damage observed at the top of the dam leads to the conclusion that the dam moved violently during the earthquake, especially at the central region (monoliths 20, 21, 22, 23 and 24, see Fig. 28) where the dam is tallest. It is likely that the peak accelerations may have reached 0.6—0.7 g as inferred from the strong-motion data.

During subsequent years, the dam has been strengthened considerably and the project engineers are confident that the dam is now strong enough to withstand earthquakes of a magnitude and intensity comparable to that of the Koyna earthquake of December 10, 1967.

Chapter 3

SEISMIC RESERVOIR SITES: THEIR GEOLOGY AND SEISMICITY

More than 20 examples in several countries are now known where earthquakes have been initiated or intensified following the impounding of large
artificial reservoirs (Rothé, 1968, 1969, 1970; Gupta et al., 1972a, National
Academy of Sciences, U.S.A., 1972). Association of earthquakes with an
artificial reservoir was pointed out first by Carder (1945) for Lake Mead in
the United States. The phenomenon of reservoir-associated earthquakes drew
worldwide attention of the scientific community after the occurrence of
damaging earthquakes of magnitude 6 or more (Richter scale) at Kariba in
the Zambia—Rhodesia border region, at Kremasta in Greece, and at Koyna in
India. Fluid injection in a deep disposal well at Denver, Colorado, demonstrated that earthquakes can be triggered by increased fluid pressure in the
substratum. This phenomenon was later confirmed by the experiments of
fluid injection and withdrawal at the Rangely oil field in the United States.

In the case of Lake Mead, a previously aseismic area, thousands of local
earthquakes have been recorded very close to the reservoir since its impounding in 1935. The activity continued to increase with a maximum in 1939
which included an earthquake of magnitude 5. In some cases, such as at
Kremasta, the increase in the frequency of the earth tremors near the reservoir immediately after the impounding has been spectacular. The Kariba area
is generally considered aseismic, with the exception of a suspected active
fault along which at least one shock was felt on July 15, 1956. A burst of
earth tremors occurred in the Kariba region following the impounding of the
world's largest artificial lake. In the case of Koyna, where only a few earthquakes have been reported in the historical past from the nearby regions,
filling of the reservoir was followed by spectacular local seismicity.

The phenomenon of induced seismicity due to reservoir impounding has
been clearly demonstrated by the examples of the Hendrik Verwoerd reservoir in South Africa (Green, 1974) and the Talbingo reservoir in Australia
(Muirhead et al., 1973). In these aseismic locations, seismograph networks
were established well before reservoir impounding to monitor any change in
seismic activity after impounding. Earthquakes indeed started immediately
after the impounding began. In addition to the above-mentioned examples,
the seismic effects of reservoir impounding have been also very well documented in the cases of Monteynard and Grandval in France, Nourek in the
U.S.S.R., Oued Fodda in Algeria, Vajont and Pieve di Cadore in Italy, Marathon in Greece, Canalles and Camarillas in Spain, and other cases. However,
at some other places, such as Mangla in Pakistan and Lake Benmore in New

Zealand, reported increases in local earthquakes cannot be established without doubt; the increase may be partly due to the improvement in the detection capabilities of recording networks (Adams and Ahmed, 1969; Adams, 1974). It is also likely that in some other cases the increase in seismic activity may not really be as spectacular as reported, specifically where reservoirs are impounded in areas which have a seismic history. Moreover, most of the reservoirs have been constructed in relatively sparsely populated areas and only a few of them had been instrumented to detect seismic activity.

The triggering of earthquakes due to reservoirs can be proved or disproved by the comparison of seismic records for the periods prior to, during, and following loading. However, the fact remains that seismic records for the period prior to loading are not available in most of the cases. The matter is further confused by the fact that seismicity appears to increase in any area when sensitive seismographs are introduced. This was noticed following the installation of the World Wide Standard Seismograph Network and has been observed particularly on a more local scale when an array of seismographs was set up in the Indian Shield. The reason for such observations is simply that an area where no shocks were felt (and thus regarded as aseismic) may have small events which are detected only when instrumental records are made. There are cases where newspaper reports and other historic records have been forwarded and interpreted to construe the seismic status of a region. However, such efforts have not yielded reliable results because of the exaggerations in reporting. This is particularly true for regions which were known to be aseismic and where earthquakes occurred following artificial lake impoundings or fluid injections. The Koyna and Denver regions are typical examples. However, the seismic histories developed in this manner have been rather vague and could not exclude the possible triggering of earthquakes by reservoir impounding.

As mentioned earlier, most of the reservoir sites were rather thinly populated before the starting of the projects. However, project personnel increased the population in these areas and this factor is likely to effect the number of felt-earthquake reports. However, in many of the cases considered, such as Koyna, Denver and Lake Mead, the nearby areas were well populated. Seismograph stations were also being operated at sites close enough to record moderate-magnitude events. In the case of the Koyna area, some 50 earthquakes of magnitude ⩾4.0 have occurred since the reservoir was impounded. If any shock of even this magnitude had occurred prior to impounding, it would have certainly been felt in the area and it would have been recorded by the sensitive Benioff seismometer located about 100 km away at Poona, which had been operating for a period of 12 years prior to impounding. Moreover, there is no felt-earthquake report in the Koyna area in the 100-year period prior to the impoundment of the lake.

It has been observed by Rothé (1968) that invariably reservoirs showing seismic effects have depths of 100 m or more. These reservoirs also have large capacities, being of the order of $1,000 \times 10^6$ m^3 or more. It has now become conventional to call a reservoir large when the total volume of the water contained in it exceeds $1,000 \times 10^6$ m^3 or the maximum depth exceeds 100 m (National Academy of Sciences, U.S.A., 1972; UNESCO Working Group on Seismic Phenomena Associated with Large Reservoirs, 1973). In most of these cases earthquakes have been centered near the reservoirs. In some cases, the activity spreads over a larger area, such as at Mangla. The frequency of the earthquakes generally increases with the rise in water level and volume. But these are not the exclusive governing factors because there are many high dams and large reservoirs which do not exhibit induced seismicity. The common factors for all cases of induced seismicity seem to be the presence of specific geological conditions and the tectonic setting. The presence of pre-existing slip planes seems to be a necessary prerequisite.

In the following, a review of the geology, seismicity, and the relationship between water level and earthquake frequency for these cases will be presented. The most significant cases, i.e. Koyna, Kariba, Kremasta and Lake Mead, have been considered in detail. The hydrology of Koyna, Kariba and Kremasta areas has been discussed in order to assess the seepage flow of water to the basement rocks.

KOYNA DAM, INDIA

Geology

The Shivaji Sagar Lake is filled by the Koyna River which flows due south for about 60 km cutting a 500—600 m deep N—S-trending valley through the Deccan Traps until it takes a sharp easterly turn near Helwak and then flows another 60 km before joining the Kirshna River (see Fig. 1).

The Koyna Dam is a concrete gravity dam of 853 m length and 103 m height. The reservoir has a capacity of $2,780 \times 10^6$ m^3 with a maximum water depth of 100 m. The dam is founded on Deccan Traps (Fig. 36). In the area covered by the Deccan Traps, two distinct geomorphic patterns are seen, first, the high plateau of the Sahyadri Mountains, limited to the west by the continental divide forming a magnificient scarp; and second, a gently rolling country almost at sea level, from the west of the continental divide to the coast, which is known as the Konkan. The average level of the Deccan Plateau, which extends for large areas in the interior of the Peninsula, is 600 m; the peaks of the escarpment, which are erosional remnants, rise to heights of 1,650 m. South of 20° N latitude, the continental divide lies within 100 km of the western coast. The drainage pattern in the plateau area,

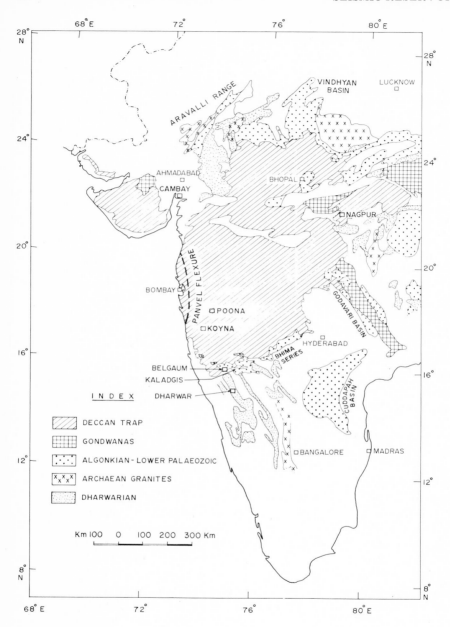

Fig. 36. Simplified geological map of Peninsular India.

except the N—S course of the Koyna River, is towards the east with broad alluvium-covered valleys.

The geology of the Koyna and the region surrounding it has been investigated by many geologists, for example, Krishnan (1960), the Geological Survey of India (1968), Committee of Experts (1968), and others. We review the results of these investigations in the following paragraphs. Since the basement rock cannot be directly observed due to the thick cover of the Deccan Traps, the salient geological features in the surrounding Peninsular Shield which may possibly extend beneath the traps, are described in the following.

Archaean and Upper Precambrian formations are exposed over more than half of the Peninsular Shield of India. The rest of the area is occupied by Gondwana and later sediments and by the Deccan Traps (Fig. 36). The major mountain-building activity ceased in pre-Vindhyan (Precambrian) times; however, minor folding, block faulting and epeirogenic movements have continued in the later epochs.

The oldest-known formations, south of the Deccan Traps, are the Archaean rocks of the Dharwar System consisting of schists and gneisses, which form the basement in the region. The Dharwars have a general NNW—SSE trend and are surrounded by Archaean gneisses and granites. No major faults are known to exist within the Dharwar rocks.

The close of the Archaean era was marked by strong folding due to intense orogenic forces and granitic intrusions. This was followed by a prolonged period of denudations, and then by the Purana era during which the Cuddapah and Vindhyan Systems belonging to Algonkian and Cambrian times were deposited in various parts of the Peninsula. In this area, resting directly on the weathered surface of the Dharwarian rocks, are the Algonkian Kaladgis, consisting of basal conglomerates, sandstones, hornstones, and shales. There is an interface of Kaladgis and traps about 70 km south of Koyna and several inliers of Kaladgis occur surrounded by Deccan Traps. East of the Kaladgis are the rocks of the Bhima Series which show evidence of disturbance at the junctions with the Deccan Traps. These rocks correspond to the Vindhyan System and consist of sandstones, shales and limestones.

After the deposition of the Vindhyan rocks and their uplift, there was a great hiatus in the stratigraphic history of the Peninsula. However, in Carboniferous time the Precambrian Shield was subjected to severe tectonic stresses, causing a series of fractures and grabens. Later, the Gondwana sediments from Upper Carboniferous to Jurassic or Middle Cretaceous times filled these grabens. The Gondwana sediments are thick and are of freshwater origin. Several inliers of Gondwanas occur in the northeastern portion of the Deccan Traps.

Beneath the northern and eastern tracts of the Deccan Traps, thin layers (up to 20 m) of marine and fluviatile sedimentary rocks of Late Cretaceous

age overlie various older formations such as the Archaeans and the Gondwanas. Along the Godavari River, the infratrappeans consist of marine sandstones. Because of the thin development and the uneven surface of older
formations, these sediments do not form a coherent cover and, therefore, the
lower basalt flows rest not only on these sediments, but also on older
Gondwana sediments and Precambrian rocks.

The NNW—SSE-trending Dharwars and NW—SE-trending Gondwana rocks
of the Godavari Basin, after erosion during Late Mesozoic time, may be
aligned with a general NW—SE trend below the lavas in the western part of
the Deccan Traps. In the Aravalli Range (Archaean) of Rajasthan, there is a
major NE—SW system of faulting which may extend below the alluvium to
the Gulf of Cambay. This fault is old but might have been rejuvenated
during Mesozoic or Tertiary time.

Volcanics

From the uppermost Cretaceous to Eocene times (Krishnan, 1960;
p. 485), a series of basaltic lava flows erupted subaerially from numerous
fractures in the crust. Called the Deccan Traps, they are one of the largest
flood eruptions of the world, covering over a half million square kilometers
of the western Peninsula (Fig. 36).

The lava flows spread out as nearly horizontal sheets, the earliest flows
filling up the irregular topography which was an erosional surface created in
Late Mesozoic times. Between Belgaum and the western coast, the base of
the volcanics descends from an elevation of 800 m to sea level within a
distance of 70 km.

The flows are called traps in view of their terrace-like appearance. The
present topography is the result of activity in Late Tertiary and Pleistocene
times followed immediately by the fluviatile and eolian activities of Recent
times. The volcanics have a thickness of at least 2,000 m in the Konkan and
Western Ghats just east of Bombay, and thin out towards the east, forming
only a thin layer around Nagpur. A great scarp of the Western Ghats exposes
a 1,300-m section of the lava flows and sedimentary intertrappean beds.

Folds, faults and fractures

Over much of the area, the Deccan Traps are virtually horizontal. In the
high plateau, the flows exhibit a very gentle easterly dip of 1°. In the low
country of the Konkan the flows show a westerly dip of 3° to 4° or more
(Das and Ray, 1972). North of Bombay, along the coast, it has been observed that these lava flows dip up to 15° in a westerly direction. This
feature with the western limb dipping to the west, is called the Panvel
Flexure, in view of its monoclinal structure. In the Koyna area a westerly dip
of 1° has been noted (V.S. Krishnaswamy, personal communication, 1973).

Major faults have been rarely observed in the Deccan Traps. Some geologists believe that the great scarp of the Western Ghats extending 450 km in an N—S direction is the result of a fault; however, it may be due to erosion (Committee of Experts, 1968). The N—S trend of 33 hot springs between Surat and Ratnagiri (Fig. 37) strongly supports the existence of such a fault. Geophysical investigations, which will be discussed later, also indicate the existence of a basement fault parallel to the west coast in this region. A N—S satellite fault along the western bank of the Koyna River has been indicated by gravity profiles. Focal-mechanism studies for the Koyna earthquake (Chapter 4) indicate a left-lateral strike-slip movement along a vertical N—S-trending fault. The straight west coast from Cambay to the southern tip of India has been generally assumed to be the result of a major fault during Late Pliocene to Early Pleistocene times, but so far no definite evidence, either on land or in the offshore areas, has been found. Minor epeirogenic movements have occurred along the west coast from Pliocene to Recent times. The Cambay Basin is a N—S graben in which the lavas were faulted

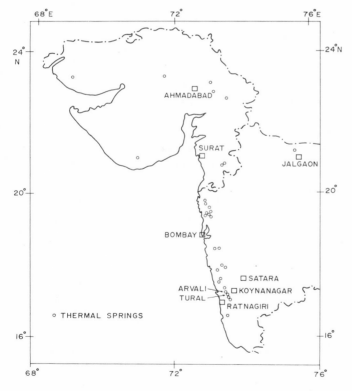

Fig. 37. Location of the thermal springs in Western India (adapted from Geological Survey of India, 1968).

between Late Cretaceous and Eocene time. They occur at a depth of 2,000 m below sea level with eroded volcanics and Tertiary sediments overlying them.

Much of the Peninsula is affected by a series of fractures along which brecciation and shattering has taken place, although displacement is not evident. These are believed to be due to shearing of the crust. Similar fractures are present in the Deccan Traps with an orientation between NNE and WNW. These fractures are younger in age than the dykes and may be of Early Miocene age. The fractures are abundant along the broad crestal part of the Panvel Flexure in the Konkan, north-northeast of Bombay. Across the Tapti River, near Vajpur (21°21'N, 73°46'E), the fractures are abundant but show no lateral displacement of the dykes. Das and Ray (1972) have indicated that the fractures along the Panvel Flexure, which are aligned from NNW to NNE, are of the nature of shear and tension joints as distinct from the normal cooling joints of the basaltic lava. The rocks of the sheared zones reveal a shattered nature; in some places slickensides indicate movement along the shear planes. These fractures are believed to be caused by secondary stresses from the reactivation of the suspected basement fault. In the northern and western part of the Koyna catchment, the fractures are aligned from NNW to NNE and are up to 20 km in length, cutting at least 800 m of volcanics without any displacement. In the tail-race tunnel of stage III of the Koyna Project, a group of such fractures with orientation from N to NNW is well displayed. The fracture zones have a width varying from 1 to 20 m, in which the rocks are shattered and form clay gouges. The N—S stretch of the 500—600 m deep Koyna Valley which is against the general slope of the country, may be due to a fault, or more likely, to fractures. Similarly, the eastward turn of the river from a mainly N—S course has been interpreted as the result of a major E—W fracture (Fig. 1).

Gravity anomalies, refraction profiles and hot springs

Peninsular India is characterized by negative Bouguer anomalies with an average value of —60 mgals. In the Koyna region the anomaly is of the order of —100 mgals (Bhaskar Rao et al., 1969, Kailasam et al., 1972). This negative Bouguer anomaly could be interpreted in many ways. Kailasam et al. (1972) have attributed it to a synclinal sag. According to Guha et al. (1974), it could be due to a buried anticline of low-density granite or Kaladgis (average density of granites and Kaladgis is 2.7 compared to 2.9 for the basalts). Brahmam and Negi (1973) explain that it is due to a rift valley which developed prior to the deposition during the Precambrian and possibly accommodated the Precambrian formations and the Gondwana sediments. Fig. 38 shows the Bouguer gravity anomaly values along a roughly E—W regional traverse from Pophli (west of Koyna) to Guhagar near the west coast (Kailasam and Murthy, 1969). There is indication of a fault near Ram-

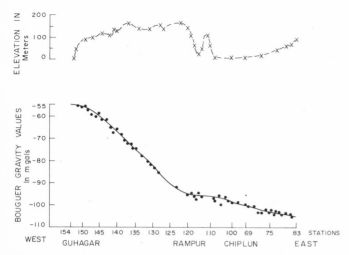

Fig. 38. Bouguer gravity anomaly along a roughly E—W regional traverse from the Koyna area to the west coast of India (after Kailasam and Murthy, 1969).

pur (17°29'N, 73°25'E; Fig. 38), about 18 km east of the sea coast. Kailasam et al. (1969) have also reported deep refraction soundings at Karad (17°17'N, 74°11'E), Pophli (17°28'N, 73°38'E), and near Rampur (17°29'N, 73°25'E). These indicated discontinuities with a higher-velocity layer at depths (from the ground surface) of 500 m near Karad, 200 m at Pophli, and about 670 m near Rampur. These data also support the existence

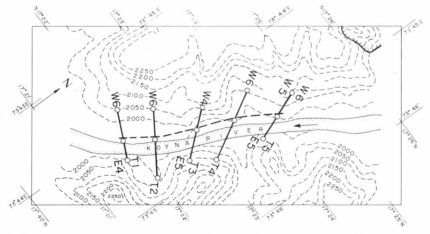

Fig. 39. Locations of the gravity and magnetic profiles across the Koyna River (after Kailasam and Murthy, 1969). The heavy dashed line indicates the axis of the gravity anomaly.

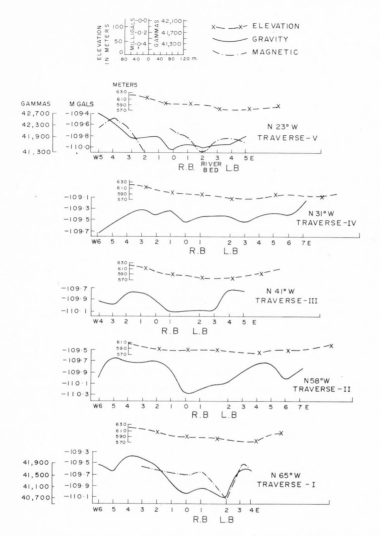

Fig. 40. Bouguer gravity anomalies and magnetic anomalies along the profiles across the Koyna River (after Kailasam and Murthy, 1969).

of a fault near Rampur, as suggested by the regional gravity traverse. The N—S-trending zone of hot springs (Fig. 37), where 33 hot springs have been located between Surat and Ratnagiri, further supports the presence of the suggested fault near Rampur.

Another set of gravity and magnetic profiles has been carried out by Kailasam and Murthy (1969) across the Koyna River to the south of the Koyna Dam just before the easterly turn of the river (Fig. 39). A sharp drop in Bouguer gravity anomalies is noticeable around the stations denoted by open circles along all the 5 profiles, which are located on the western bank of the Koyna River (Fig. 40). On the basis of these data, they suggested the existence of a fault in a roughly N—S direction along the river course. It is believed that this fault extends down to the basement rock.

Hydrology

The basalt flows have a widely varying textural character, which may be massive, amygdular or vesicular. The vesicular and non-vesicular flows alternate with each other and many times these flows are separated by ash beds, red-bole beds, or lacustrine sediments which are called intertrappeans. In some places, they also alternate with the massive flows. The thickness of the individual flow varies from a few meters to 40 m. Numerous ash beds occur in the Upper Traps around Bombay, Poona, and in the Western Ghats. This is due to eruptions of a violent nature which resulted in the formation of the high Sahyadri Mountains. Ash beds are usually brecciated with the fragments of trappean basalts embedded in uniformly dusty material. The breaks in the successive flows are marked by red-bole horizons and intertrappean beds. Red bole, which may be old lateritic soil, forms a highly clayey horizon occurring between the massive and underlying vesicular traps.

The freshwater intertrappean sedimentary beds are intercalated with both the lower and upper sections of the traps. These beds are fossiliferous and are found to be of Early Tertiary age. The ratio of sedimentary to basaltic rocks is 1 : 10. The lower intertrappean beds are of small horizontal extent and a thickness rarely exceeding 6 m. They consist of cherts, impure limestones, and pyroclastic materials. The upper intertrappean beds confined to the west coast around Bombay are much thicker (about 30 m), and consist of shales (Krishnan, 1960, p. 482).

The trap rocks are poor producers of water and have lesser water-holding capacities than sand and gravel beds. The water-holding capacity is further reduced by their horizontal dips and steeply eroded hill slopes. Wherever lateral truncation of the trap flows due to erosion occurs, groundwater from these flows seeps out through the sides of their slopes into the gullies, streams and rivers. The basalt flows are drained so rapidly, that by summer the wells become useless for irrigation purposes as a result of the rapidly lowering water table.

The groundwater possibilities in the Deccan Traps are to a great extent governed by the nature and constitution of the individual flows. The massive traps with their fracture porosities, the vesicular traps with their interconnected and partly filled vesicles, and the intertrappean sediments with their primary porosities provide a good yield of water.

In the upper traps of the Western Ghats and Poona region, ash beds and intertrappeans are prevalent. The traps are characterized by a highly erratic behavior and thin out within short distances caused by an explosive phase of volcanism. The intertrappean sediments in these regions have provided good yields of water. The dolerite dykes, which are present especially in the Konkan, control the movement of water. Extensive fissures in the area possibly act as feeders.

East of the Poona region in the Sholapur district, the traps are characterized by areally extensive units which can be traced laterally for tens of kilometers. These originated as a result of a slow and quiescent type of extrusion of lava floods along linear fractures. In the Sholapur district, the vesicular traps are thick and areally extensive, dipping gently towards the east. They have artesian conditions in the areas in a down-dip direction. Groundwater in them occurs under water-table condition in the vesicles and cavities. These units yield large quantities of water.

The extensive fractures and fissures in the Koyna area possibly act as feeders for the trappean flows and they facilitate the deep circulation of water. The deep circulation of water is evidenced by the presence of hot springs in the west coast. The hot springs are related to fractures which probably resulted from the dislocation of the Deccan Trap platform during Late Tertiary times. The maximum temperature of these hot springs is about 70°C (Chaterji, 1969), and assuming a geothermal gradient of about 25°C/km, a surface temperature of 25°C and allowing for the heat losses, a depth of 2 km for the water circulation has been inferred (M.L. Gupta, personal communication, 1974), assuming that tectonism and magmatism do not contribute appreciably.

Evidence for an increased discharge in the springs after reservoir filling is not conclusive; however, in the Tural Spring (Fig. 37), about 25 km west-southwest of Koyna, the discharge measured after filling of the reservoir is 10.5 liter/sec (Gupta and Sukhija, 1974), which is about 10 times more than the pre-filling value of 1.3 liter/sec (Chaterji, 1969). Another spring at Arvali, situated close to the Tural Spring shows the same order of discharge before and after reservoir filling.

Vesicular and amygdular flows along with broken and sheared zones are encountered at the dam site as seen in the geological section shown in Fig. 27. In the tail-race tunnel, numerous vesicular flows and sheared and jointed sections are encountered from which water continues to ooze.

Seismicity

The Koyna Dam (17°24'N, 73°45'E) is situated in the Peninsular Shield of India, which is considered to be one of the world's geologically stable and nearly non-seismic Precambrian Shields. No earthquake of appreciable intensity has been experienced during the last few decades in the shield area, and hence it has been regarded as aseismic. Historical records, however, show that during the last 600 years some earthquakes of moderate intensity and several tremors were experienced in the Peninsular Shield, particularly along the west coast conforming with the seismic activity along the marginal areas of the stable shields (De Sitter, 1965, p. 479). The seismic activity along the west coast may also be due to the possible extension of the trench developing in the Indian Ocean between Ceylon and Australia up to the western coast of India (Sykes, 1970). A catalogue has been prepared by Guha et al. (1970) for earthquakes which occurred in Peninsular India and its neighborhood in the period from 1341 to 1969. Gubin (1969) has prepared a map showing the epicenters in the Peninsular Shield (Fig. 41). During the last few centuries, the earthquakes at Bellary (1843), Coimbatore (1900), Koyna (1967), Bhadrachalam (1969) and Broach (1970) have been the only known significant seismic events of the shield area, the Koyna earthquake of December 10, 1967, being the most severe.

Prior to the impoundment of the Shivaji Sagar Lake in 1962, there was no

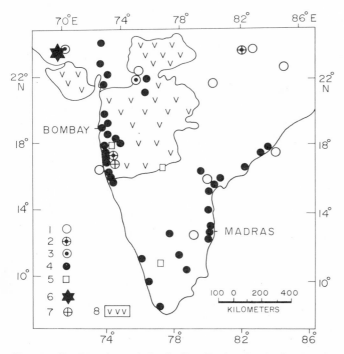

Fig. 41. Earthquakes of the Indian Peninsula. *1—3* are epicenters determined instrumentally: *1* = earthquakes, M = 5 to 6; *2* = earthquake of Son Valley, 1927, M = 6½, intensity 7, and earthquake in the Western Ghats, September 13, 1967, M = 5—5½, intensity 6—7; *3* = the Rann of Cutch earthquake of 1956, M = 7, intensity about 9, and Satpura earthquake of 1938, M = 6¼, intensity 8; *4* = the point at which a local tremor of intensity 5 or 6 was recorded; *5* = the central point of intensity 7 area of a local shock; *6* = the probable center of the area where in 1819 an earthquake of intensity 10 occurred; *7* = the Koyna earthquake, 1967; *8* = Deccan Traps (adapted from Gubin, 1969).

seismic station operating in the area, and hence no instrumental record is available of possible weak tremors. However, many aged persons interrogated during visits to several villages within a radius of 50 km from the lake categorically denied their experiencing any sounds or tremors prior to 1962. Engineers, engaged in the construction of the dam, also reported no occurrence of earthquakes. Scrutiny of the records of the Benioff seismometers operating since 1950 at the Poona seismographic station, located at a distance of 115 km north of the dam, revealed no significant tremors which could correspond to this area.

After filling started in 1962, mild tremors accompanied with sounds similar to blasting began to be prevalent and the frequency and intensity of

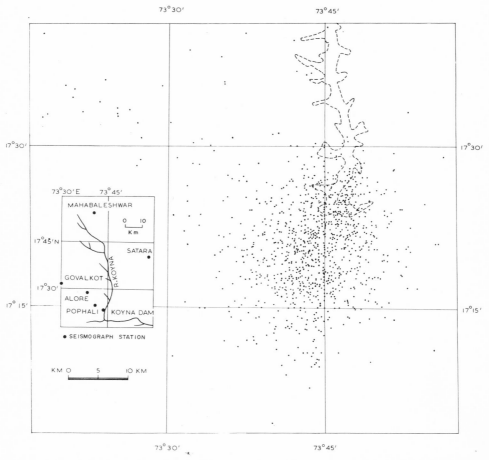

Fig. 42. Epicenters of the earthquakes which occurred during the period December 10, 1967, to December 31, 1971, around the Koyna reservoir (after Guha et al., 1974).

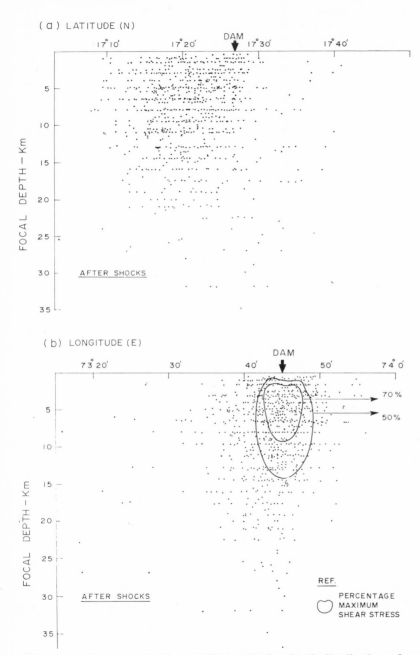

Fig. 43. (a) Latitude—depth, and (b) longitude—depth distribution of earthquake foci in the Koyna area from December 10, 1967, to December 31, 1971 (after Guha et al., 1974).

tremors, especially near the dam site, increased considerably from the middle of 1963 onwards. Intermittently, the strongest of these tremors would rattle the windows, disturb utensils, etc., with a maximum intensity of V (MM scale). Installation of seismographs and some other instruments, such as accelerographs, tiltmeters, etc., at the dam site and around the lake area began from the end of 1963.

By the end of 1969, seven seismological observatories were established around the reservoir, including two at the Koyna Dam (inset Fig. 42). One observatory along a gallery at the Koyna Dam was set up in late 1963 and three other observatories, at Satara, Mahabaleshwar and Govalkot, were established a year later equipped with Wood-Anderson seismometers. The Benioff seismometers were commissioned at these observatories in January 1965. Three more observatories were later established at the Koyna Dam (downstream), Pophli and Alore. The two Koyna Dam observatories are also equipped with accelerographs and tiltgraphs.

Guha et al. (1968) located the foci of the tremors down to a depth of 4 km. These shocks were believed to be due to reservoir settlement. The unexpected earthquake on September 13, 1967, of magnitude >5 in the Koyna area caused mild damage to some of the buildings at Koyna and the underground power plant. This and the December 10, 1967 earthquake, which had a magnitude of 6 and which caused considerable amount of damage, were thought by many earth scientists to be unrelated to the reservoir (Committee of Experts, 1968).

The epicenters of the earthquakes which occurred before December, 1971, are plotted in Fig. 42. Most of the epicenters lie within 25 km of the dam and are mostly situated downstream of the dam. Focal-depth determination, though not very accurate, show depths reaching about 30 km for some earthquakes, but most earthquakes have their hypocenters within a 15 km depth (Fig. 43a,b). The foci of a majority of the earthquakes (Fig. 43b) lie within a 50% contour of the maximum shear stress due to the water load (Guha et al., 1974).

Reservoir level and earthquake frequency

The correlation of the water level in the Koyna reservoir and the frequency of shocks near it was pointed out by Gupta et al. (1969). As mentioned earlier, there was a near-absence of seismic activity in the region prior to filling. A careful examination of Fig. 44 indicates that every year, following the rainy season, the seismic activity increases. The data for the five rainy seasons from 1963 to 1967 provide five such examples. Of these five examples, the following two are very conspicuous. The highest water level for the longest duration was retained during August to December, 1967, and this corresponds well with the maximum seismic activity including the magnitude 5.5 (Richter scale) earthquake of September 13, 1967, and the magnitude

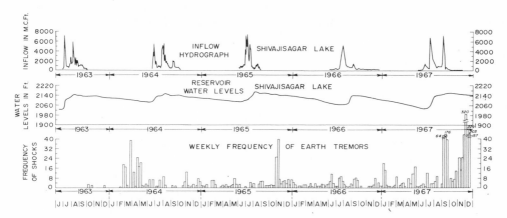

Fig. 44. Inflow hydrograph and water level of the Koyna reservoir compared to the seismic activity in the area (after Guha et al., 1968).

6.0 earthquake of December 10, 1967. The second highest level was reached during August to October, 1965, which corresponds with the second most conspicuous activity in November 1965. The slow rate of loading and the comparatively lower levels following the rainy seasons in 1964 and 1966 were characterized by relatively less conspicuous seismic activities with larger time lags. In general, the seismic activity follows reservoir loading with a certain lag of time.

After 1967, the reservoir levels in the Koyna reservoir have been relatively much lower. It is also noticeable that no significant earthquake occurred in the region since the magnitude 5.0 earthquake of October 29, 1968. However, due to good rains during 1973 and the absence of seismic activity during the preceding 5 years, the reservoir was permitted to be filled to its maximum capacity on August 15, 1973. This was followed by a conspicuous increase in seismic activity, including an earthquake of magnitude 5.1 which occurred on October 17, 1973.

The seismological station of the National Geophysical Research Institute (N.G.R.I.) at Hyderabad is situated on a very quiet foundation, and is equipped with sensitive Benioff seismographs, which have a magnification of 10^5. As such all earthquakes of magnitude $\geqslant 3$ from the Koyna region at an epicentral distance of 490 km are distinctly recorded at this station. An analysis of the records of this observatory has shown a decrease in the seismic activity of the Koyna region (Gupta and Rastogi, 1974a). The list of earthquakes with a magnitude $\geqslant 4.0$ in the Koyna region has been updated and reproduced in Table VII. As can be seen from this table, the frequency of the earthquakes of magnitude $\geqslant 4$ has decreased during the last few years after the damaging earthquake of December, 1967. Nine such earthquakes occurred during 1969, seven during 1970, four in 1971 and three in 1972.

TABLE VII

Koyna events of magnitude ≥ 4 since 1969

Date	Magnitude	Date	Magnitude
January 21, 1969	4.1	January 23, 1971	4.2
February 13, 1969	4.2	February 14, 1971	4.0
March 7, 1969	4.4	August 10, 1971	4.0
June 3, 1969	4.2	August 10, 1971	4.3
June 27, 1969	4.5	May 1, 1972	4.2
July 22, 1969	4.0	May 11, 1972	4.5
November 3, 1969	4.1	November 11, 1972	4.1
November 4, 1969	4.2	April 19, 1973	4.1
April 16, 1970	4.0	October 17, 1973	4.0
May 27, 1970	4.8	October 17, 1973	4.1
June 8, 1970	4.1	October 17, 1973	5.1
June 17, 1970	4.1	October 24, 1973	4.6
September 21, 1970	4.0	November 11, 1973	4.6
September 25, 1970	4.6	February 17, 1974	4.5
September 26, 1970	4.6		

During 1973 also, before October, only one earthquake of magnitude 4.1 occurred, on April 19. However, in October 1973, in addition to the magnitude 5.1 earthquake, three earthquakes of a magnitude exceeding 4 occurred. These statistics show that the spurt of activity during late 1973 does not agree with the trend of gradually decaying seismicity in the Koyna region from 1968 to the middle of 1973. This very strongly supports the persistent relationship between water level and earthquake activity. After the October 17 earthquake, no earthquake as strong occurred during the remainder of 1973, probably due to the fact that the duration for which high levels have been retained is much less compared to 1967. During 1967, a water level above 2,144 ft (653 m) was maintained for 132 days, while during 1973 this level was maintained for only 98 days. The correlation coefficients between reservoir level and tremor frequency has been computed and is discussed later.

LAKE KARIBA, ZAMBIA

Geology

The Kariba Dam is a double-curvature arch dam having a height of 128 m and length of 617 m. This dam forms the largest artificial reservoir, having a capacity of $175,000 \times 10^6$ m^3. Geological investigations in the Lake Kariba area have been carried out by many workers, e.g. Bond (1953, 1960), Hitchon (1958), Knill and Jones (1965), Drysdall and Weller (1966) and Kirk-patrick and Robertson (1968). Snow (1974) summarized the geological aspects pertinent to the hydrology, loading and seismicity. These aspects are mentioned here.

According to Walters (1971, p. 351), the following is the main sequence of strata in the region.

Triassic	Upper Karoo sandstone	
	escarpment grit and conglomerate	} Karoo Sequence
Permian	Madumabisa argillaceous mudstone	
Precambrian	quartzite	
Archaean	gneiss, amphibolites, schists	

Lake Kariba is situated in a NE-trending trough (Fig. 45). Zambia and

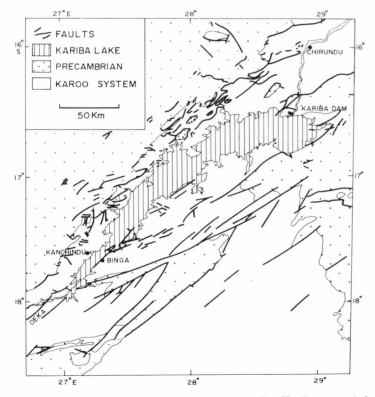

Fig. 45. Geology and fracture pattern in the Kariba area (after Gough and Gough, 1970b).

Rhodesia are situated northwest and southeast respectively of the NE-flow-
ing Zambezi River. The Karoo Series of Carboniferous to Triassic age constit-
utes the bedrock. The Precambrian basement has been locally exposed at a
number of places such as at the dam site, by the Zambezi River. The basement complex around the Zambezi trough generally consists of coarse-

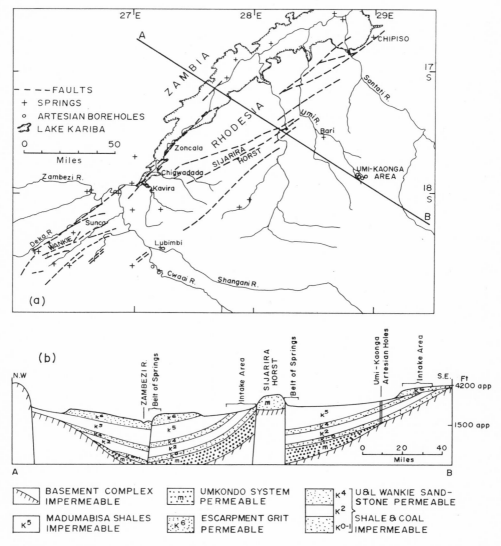

Fig. 46. (a) Map of the Middle Zambezi Basin showing the thermal springs and faults. (b)
Section across the Zambezi Valley showing the artesian structures which control the hot
springs (after Bond, 1953).

grained gneissic rocks with steep foliation striking NE. Exposed rocks are highly fractured and faulted. At Kariba, along the Zambezi Gorge, the gneissic rocks are cut by Precambrian pegmatites and Jurassic dolerites. The Precambrian foliation, pre-Karoo rift valley, and post-Karoo (Cretaceous) faults and folds all follow the NE trend. An unmetamorphosed series of Precambrian rocks, lying unconformably over the gneisses, was faulted during the pre-Karoo rifting, and may be buried beneath parts of the Karoo Trough (Bond, 1953). It is called the Sijarira Series and is exposed southwest of Lake Kariba in Rhodesia (Fig. 46a,b). It is predominantly composed of quartzitic sandstone with interbedded shales, and its total thickness exceeds 600 m. In East Africa, several grabens were formed in pre-Karoo times by processes similar to the large-scale Tertiary rifting. Before the Karoo deposition, the basement was uneven. Jurassic and Cretaceous faulting enhanced the basement irregularities. Deposition of the Karoo sediments started in downwarped troughs which were precursers to rift-like structures.

The Karoo Sequence is more than 2,800 m thick. The basal beds are Dwyka tillite and outwash gravels at some places, and conglomerates and sandstones at other places. Above the basal beds are Ecca coal beds and sandstones which have a combined thickness of 100 m. Above these lie the 450—640 m thick Madumabisa mudstone of Permian age. The Madumabisa rocks are overlain by 365 m of coarse sandstones of Triassic age, which are known as the escarpment grit as they form a prominent escarpment (Fig. 46a, b). Above this is the Stormberg Series consisting of 120 m of pebbly arkoses, mudstones, and basaltic lavas.

The Karoo Sequence was disrupted in Jurassic and Cretaceous times by numerous NE-trending normal faults which dip 50—65° in both SE and NW directions. The faults are exposed in the Momba Colliery on the northwest shore of Lake Kariba and in the Wanki coal mines of southern Rhodesia. During the Miocene, Pliocene, and Pleistocene, these faults were reactivated in some basins other than the Mid-Zambezi; the current seismicity at Lake Kariba apparently seems to be related to this fault system. A distinct fault contact exists at the northwest flank of the asymmetric Mid-Zambezi Trough. Graben and horst structures were formed as a result of the later breaking of the southeast flank which was originally a monoclinal flexure. The Sijarira block, east of Binga, is a prominent horst (Fig. 46a, b).

Continental crustal movement took place during the breakup of Gondwanaland during Middle Jurassic to Early Cretaceous times. The rift structures that localized the Permo-Carboniferous Karoo beds seem to have controlled the Cretaceous sedimentation as well. Cretaceous sediments are seen on the Rhodesian side of the Kariba Gorge and at Kanchindu on the Zambian shore of Lake Kariba.

The course of the Zambezi River was probably determined after the formation of rift structures and before the deposition of Cretaceous sediments. Good evidence of wrench faulting which displaces Karoo rift faults and dinosaur beds exists in Malawi. As noted by King (1962), profound rifting lowered the Karoo beds along the Mid-Zambezi Trough.

Erosion took place from Cretaceous to Miocene times. During the Miocene there is no evidence of reactivation of the Mid-Zambezi faults; however, active rifting developed elsewhere in East Africa.

During the Miocene epeirogenic movements began which uplifted the erosion surfaces and initiated a new cycle of erosion in the Kariba area and upstream of Victoria Falls. During the Pliocene, faulting took place in the Tanganyika Trough and Luanga Valley. In the Late Pliocene, the post-African surface was cut, probably due to the tilting of the older surfaces which caused rejuvenation of the Zambezi River and its tributaries. In the Mio-Pliocene, normal faulting took place in Kenya and southern Tanganyika. During the Pleistocene, active rifting took place elsewhere but the Mid-Zambezi Trough remained dormant.

Nature of the prevailing stresses

At the Deka Fault, great stresses do not seem to have prevailed normal to the fault surface when it ruptured as there is no mylonite and cataclasis of the quartz grains and there is only minimal brecciation (Mann, 1967). A downward vertical displacement of at least 300 m on the northwest side has been deduced from stratigraphic investigations.

The downslip of the northwest block along the fault plane striking NNE was suggested from the first-motion study for the main Kariba earthquake (discussed in Chapter 4). This is very similar to the movement along the Deka Fault. However, wrench-fault stresses have been suggested in the region by Mann and Bloomfield (Snow, 1974) and wrench faults have been observed in southern Malawi, as discussed earlier. These alternatives should be treated cautiously as normal faulting has been persistent during the past 300 million years, and has also been indicated from focal-mechanism studies of earthquakes.

Hydrology

Regional hydrology has been discussed in detail by Snow (1974). Most of the formations in East Africa appear to be nonaquifers, but it is suggested that a water table persists in the metamorphic rocks throughout Rhodesia and Zambia. Perennial streams are found to occur in the gorge tributary to the Zambezi River. The joints and faults seen in the gorges have hydrological significance at a depth of hundreds, if not thousands, of meters. The influence of pore pressure is still great at depths exceeding 1 km, although conduits become less frequent and finer in aperture with depth. Rocks are mechanically influenced by fluid pressure in microscopic fracture, although they may not produce noticeable water in wells. Failure on surfaces conducting pore-pressure transients is critical, although it may be rare.

The Sijarira Series, believed to underly the Karoo in the Mid-Zambezi Trough, consists of coarse, clastic rocks which are not thoroughly cemented and are thus pervious. Bond (1953) has classified them as aquifers. The Karoo rocks have important hydrological significance for reservoir loading. The basal Dwyka beds are likely aquifers; Ecca coal beds are also observed to be permeable. The fine-grained Madumabisa shales may be impervious but the faults which intersect this bed are open or brecciated (Fig. 46b). The escarpment grit, in which the artesian flows have been encountered in the drill holes, is pervious. The faults in the Mid-Zambezi Trough have hydrologic significance, as evidenced by the numerous hot springs arising along them in the region south of Lake Kariba, especially near Wankie (Fig. 46a). Some prolific springs emerge along the Deka Fault which extends to Lake Kariba.

Using borehole measurements of the geothermal gradient, Bond (1953) estimated from spring temperatures the depth of meteoric water circulation in various parts of the Mid-Zambezi Basin. His interpretation of the paths of deep circulation is shown in Fig. 46b.

The hydraulic continuity from the surface to great depths in the basin is demonstrated by the hot springs. The pore pressure in the basin, when submerged by Lake Kariba, is likely to be influenced by the normal faults along which the hot springs rise. Although not permeable enough to produce a well, the fractured basement rocks in contact with the aquifers are capable of transmitting the potential changes to a few kilometers depth. This could particularly happen in dead-end, steeply inclined openings and thus propagate the surface hydraulic changes to hypocentral depths. It can therefore be visualized that reservoir loading caused regional changes of hydraulic potential which may have the crucial mechanical effect of decreasing the horizontal and vertical effective stresses in areas away from the surface reservoir.

Seismicity

Prior to the impounding of Lake Kariba, no seismological observatories existed in its vicinity and hence the exact information of the seismic status prior to the loading of the reservoir is missing. However, there is evidence of seismic activity along the fault at Binga (Fig. 45) which is near the west end of the lake (Archer and Allen, 1969). One tremor was felt near Binga on July 15, 1956 (Gough and Gough, 1970b). However, Kariba was considered to be a non-seismic area and no earthquakes for the region have been mentioned in the catalogue prepared by Gorshhov (1963). Snow (1974) reports minor seismic activity, remote from Lake Kariba, that suggests modern tectonism. However, no pre-impounding quakes occurred at the northeast end of the lake. The Deka Fault, which extends southwesterly for about 330 km from Binga to Botswana is said to be active (Mann, 1967). The Okavanga swamp

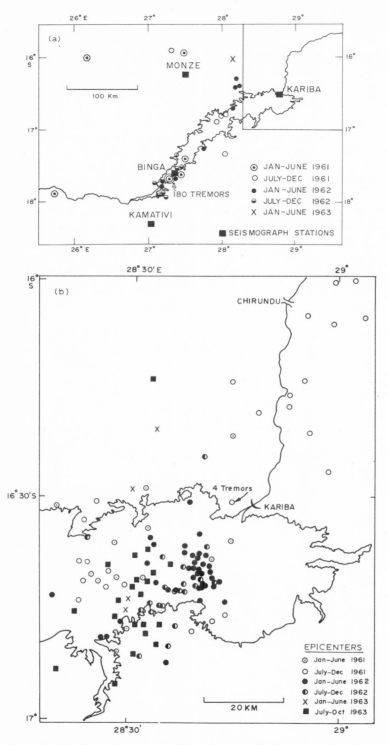

Fig. 47. (a) Epicenters west of the Sanyati Basin area and seismograph locations. (b) Epicenters in the Sanyati Basin and downstream from the dam (after Gough and Gough, 1970b).

area also experiences shocks. Occasionally, shocks occur along the E—W trend of Namwalla—Kafue—Mozobuku—Chrundu, passing just north of Lake Kariba. Another remote earthquake caused damage at the Antelope Dam, south of Bulawayo, at a distance of 385 km from Lake Kariba.

Two networks of seismograph stations were used for the study of tremors near Lake Kariba (Archer and Allen, 1969). One network consists of the main stations at Bulawayo (established February 1959), Broken Hill (April 1959, now renamed Kabwe), and Chileka (December 1962); these were equipped with Willmore three-component seismographs and are situated at a distance of a few hundred kilometers from the lake. At Bulawayo, W.W.S.S. short- and long-period seismographs have been in operation since 1963. The other network consists of four lake stations situated close to the reservoir, equipped with Willmore vertical seismographs, which were in operation from 1961 to 1963 (Fig. 47).

The lake stations recorded more than 2,000 tremors during the three years of operation. Because of the irregular operation of these stations, difficulty in time-keeping and a lack of velocity information on the crust in the region, Gough and Gough (1970b) could locate only 159 epicenters. The locations of the epicenters are shown in Fig. 47a, b. All the epicenters, except the four which lie along 16°S latitude in Fig. 47a, are within the rift valley of the Mid-Zambezi which accommodates the lake. Determination of the focal depths for these shocks, as well as for those given in Archer and Allen's (1969) catalogue, has not been possible. The precision of epicenter locations is not more than 2 km, so that the available resolution does not permit definition of the positions of active faults. 1,400 shocks of magnitude ≥2.0 are reported in Archer and Allen's catalogue for the period 1959—1968, and about 150 shocks of magnitude ≥2.5 are reported in its supplement for the period 1969—1971. In this catalogue, the epicentral locations are given from 1966 onwards.

Reservoir level and earthquake frequency

One may say that the beginning of activity in 1959 after the rapid filling coincides with the start of a network of permanent stations. However, the seismic activity from 1959 to 1962 (Fig. 48) seems to be associated with the lake, since the majority of the epicenters were located in the lake region. A definite increase in seismicity, which must be attributed to the lake, began in February, 1962, and ended in November, 1962. The maximum activity and the strongest shocks occurred in 1963 including the earthquakes of magnitude 6 which occurred in September of that year, when the lake level was at its maximum. There was a month and a half's delay between the peak reservoir level and the peak seismic activity. As shown in Fig. 48, seismicity continued up to September 1964, long after the fall of the lake level which occurred from March 1963 through February 1964. Afterwards, a low level

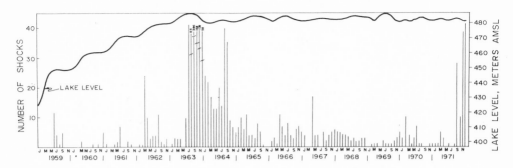

Fig. 48. Monthly tremor frequency and water levels of Lake Kariba.

of seismic activity continued until September—November, 1971, when a burst occurred, including a magnitude 3.8 earthquake. The annual fluctuations of the Lake Kariba levels have been about 2.5 m, and the burst of seismic activity followed three months after the June—July 1969 peak of 484.5 m, which was not unusual in relation to peaks of other years when the seismicity was low.

A careful examination of Fig. 48 reveals that the increased tremor frequency from April to September, 1964, corresponds well with the increase in the lake level until August 1964. Periods of reduction of water levels in October—November 1964, and during October 1965 to February 1966, were accompanied by a decline in the tremor frequency. An increased activity occurred as the lake level rose from December 1964 to June 1965, and from April to July, 1966. Sometimes the increased activity seems to be associated with the falling levels, such as those from October 1966 to January 1967 and April 1967. The statistical correlation between the Lake Kariba levels and earthquake frequency is discussed later in this chapter.

LAKE KREMASTA, GREECE

Geology

The 147 m high Kremasta Dam constructed over the Acheloos River in Greece forms a lake of $4,750 \times 10^6$ m^3 capacity. The maximum depth in the reservoir is about 120 m. Snow (1972) has summarized the geology of the Kremasta area and it has been reviewed here. The area is divided by thrust faults into three zones, viz. the Pindus, Gavrovo and Ionian zones (Fig. 49). The basement rocks in the area are unknown. In the Pindus thrust zone, east of Kremasta, siliceous limestones of Jurassic and Cretaceous age occur with ophiolites, shales and radiolarian cherts which were followed by

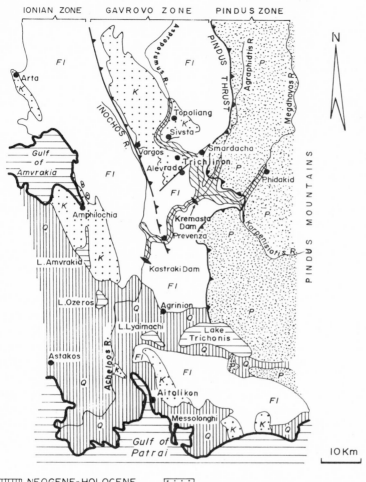

Fig. 49. Geological map of the Kremasta reservoir area (adapted from Terra-Consult, 1965).

flysch deposits (Fig. 50). The Gavrovo zone, in which Kremasta lies, is dominated by massive, gray, neritic limestones. Neritic limestones are also found in the Ionian zone, further west. Flysch deposition started first in the Pindus zone in the Upper Cretaceous. Later, flysch deposited in the Gavrovo and Ionian zones.

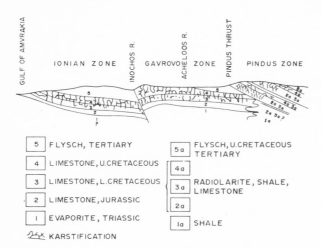

Fig. 50. Schematic cross-section from the Gulf of Amvrakia to Topoliana to the Pindus Mountains (after Snow, 1972).

In the Gavrovo limestones, important karst-solution features and considerable erosional relief developed due to uplift in Upper Cretaceous to Oligocene times. Some of the structures of these limestones are probably due to thrust faulting. Eocene subsidence to the west resulted in neritic limestones. Flysch deposition commenced during the Lower Miocene over the entire area. In the Gavrovo flysch, thick conglomerates were derived from carbonaceous and siliceous Pindus rocks. At other places, flysch consists of siltstone having a lower proportion of sandstones and lacking conglomerates. To the west of the Gavrovo zone, one or more thrust with steep eastward dip developed.

Flysch deposition was terminated during the Miocene, when the thrust movements of the Pindic Orogeny were completed and the sole of the thrust, a thick mylonite zone, developed. The Pindic thrusts probably had the incompetent halite sequence at their base as a décollement. These thrusts developed many thin plates in the incompetent shales and cherts and a few sheets in competent limestones such as those of the Gavrovo zone. The lowest Pindic thrust with pervious rocks has been submerged by the Megdhovas arm of the Kremasta reservoir. The continuity of karstic (Gavrovo) limestone across the western thrusts strongly suggests the possibility of transmission of the reservoir's hydraulic pressure to all the thrust sheets. However, due to the presence of impervious evaporites at the base of the thrusts, the transmission of the pore-pressure changes caused by the reservoir to the crystalline basement is not certain.

Folding in the area is also of Miocene age. The intensity of folding diminishes from east to west. In the Pindus zone, folding has created tight asymmetric nappes and has shattered the siliceous rocks. In the Gavrovo zone, the folds are broad and asymmetric with an NNW axis showing a dip of up to 20°. In the Ionian zone, there is no folding but a gentle ENE dip is observed. The reservoir is situated on the eastern flank of a major NNW-trending fold in the flysch rocks running parallel to the Acheloos River. The fold can also be delineated in the underlying Gavrovo limestones; however, the Gavrovo topography is probably due to exhumation rather than due to structures. Folding, subflysch relief, and faulting contribute to the exposures of limestones through the flysch cover.

The faulting took place after the Miocene folding and caused steep dipping faults trending ENE and transverse to the NNW fold axis. These faults are the youngest in the area and have displaced the fold axes dextrally. Photogeological investigations have revealed many more such faults in the unmappable and monotonous siltstones. The Pindus Thrust is cut at places by these faults. The Alevrada—Smardacha Fault belonging to this system of faults crosses the reservoir about 10 km north of the dam. Seismic activity has been confined to this area and the epicenter of the main shock is about 10 km north of Smardacha (Fig. 49). These faults, especially the Alevrada—Smardacha Fault, indicate a predominant dip-slip movement as is evidenced by the mullion structures and slickensides. The rake of 45° observed at Smardacha bridge, however, indicates equal dip and strike-slip components of motion. From the focal mechanism of the main earthquake, the dip-slip component of motion has been found to be slightly greater than the strike-slip component along a plane striking either ENE or NW (Chapter 4). Geological evidence of movement observed at Smardacha bridge supports slip along the ENE plane of the fault-plane solution. However, several steep NW faults near the Kremasta Dam, spaced about 100 m apart, have normal displacements. These faults support movement along the NW-striking plane.

According to Dewey and Bird (1970, p. 2642), the African plate is being consumed in the Ionian Trench parallel to, and west of, the fossil Aegean Trench. Mesozoic/Tertiary subduction of the African plate along the Aegean Trench has given rise to the above-described geology of the region. In the context of the plate-tectonics theory, the present process is a continuation of the Cretaceous to Miocene subduction and is dominated by E—W compression. The ENE-trending Anatoli transform fault is believed to traverse the Gulf of Patras about 50 km south of the Kremasta region. Galanopoulos (1967a), has indicated a conjugate set of wrench faults about 10 km north of the dam. The Hellenide trend of NNW faults, for example the Inachos Fault, are thrust faults. Earthquakes associated with the NNW thrust faults are shallow and are within the 7 km thick sedimentary layer, while those associated with the Anatoli trend are a little deeper. In the case of wrench faulting, the intermediate principal stress σ_2 must be vertical; NNW thrust faults may coexist with wrench faults if there are weak horizontal-sliding surfaces (e.g. Triassic halite). The deeper Kremasta earthquakes are deduced by Snow (1972), from this evidence, to be associated with ENE wrench faults.

Hydrology

Flysch outcrop occupies most of the area beneath and around the reservoir. Its hydrology plays a minor role in tectonics as it is generally impermeable. It is locally karstic with solution-carved faults and joints in calcareous conglomerates near the dam. Some prolific springs, at a distance of as much as 2.5 km downstream from the reservoir, emerge from cavernous limestone conglomerates through openings in the siltstone cover. The water-bearing Pindus limestones and radiolarites supply copious waters to the flysch which is collected on the impermeable mylonite at the base of the thrust. Along the Smardacha—Alevrada Fault, some small springs emerge because this fault acts as an impervious barrier to the downslope flow. Though flysch neither contains nor transmits large amounts of water, it maintains a high water table parallel to steep topography, and as such the reservoir can not greatly change the water table.

On the other hand, the limestones of the entire area are cavernous and maintain flat water tables. Since karstification took place earlier than flysch deposition, the Gavrovo limestones form a continuous aquifer beneath and around the reservoir, and beneath the impermeable flysch. The hydraulic continuity is apparently unbroken by post-flysch faulting and is maintained by continuous karstification. Certain proven losses of reservoir water, though leakage is negligible, suggest the raising of potentials in the carbonate aquifer. Access could either be through the outcrop areas on the Aspropotamus River or through the faults which intersect the flysch. Four piezometers installed in the limestones near the reservoir recorded fluctuations similar to those of the lake levels. One of them, at a distance of 6.5 km from the reservoir, had a potential lower by about 7 m, indicating a hydraulic gradient and a flow away from the reservoir. This observation is consistent with the recorded increase of flow at natural springs remote from the reservoir. The remaining three piezometers are located within 1 km of the reservoir and they maintained levels within 1 m of the reservoir elevation.

Since the formation of Lake Kremasta, the increased spring discharge evidences the raised hydraulic potential. Snow (1972) has cited many such examples. The temperature of some of the springs has decreased and the chemistry has also altered, reflecting the mixing of thermal and reservoir waters. The warm spring waters indicate deep circulation paths and the reservoir discharge would cause increased hydraulic potentials throughout the system. Increased potentials in the system are also indicated by springs that have been submerged by reservoirs and yet continue to discharge. Smardacha Spring is one such example. It has been the major source for the Aspropotamus River and is now submerged by Lake Kremasta by about

70 m of water. Another example is the Prevenza Spring submerged in the Kastraki reservoir. Though submerged, the discharge is still taking place as indicated by the smell of H_2S.

The buried karst is the most remarkable feature of the Kremasta geo-hydrology and it seems to play an important role in the local tectonics. Snow (1972) argued that when the reservoir raises the potentials in the karst, it must also raise the potentials in the flysch, on an average by about one half the reservoir rise. Changes of potential in the formations below the carbonates are unknown. The Pindus Thrust is submerged by the Megdhovas arm of the reservoir. This thrust may penetrate into granites and serpentinites well east of the reservoir permitting the water percolation into these basement rocks. However, earthquakes occurred due north of the reservoir and not in the east below the Pindus. It is likely that water circulates to the 4,000 m deep evaporites, which were thought to be self-sealing, bringing up NaCl to some thermal springs. Major thrust and wrench faults related to the Anatoli fault system intersect the evaporites. Although the Kremasta shocks have been very shallow (as judged by the unique rumbling and very local high intensity), the major event of February 5, 1966, was most likely in the basement (focal depth 12 ± 5 km). Fault leakage is, thus, suspected across the evaporites into the basement.

Seismicity

Lake Kremasta is situated in an area having a seismic history. As pointed out by Auden (1972), during 1953 to 1965, in an area of 87,000 km^2 around Kremasta between the latitudes 37°N and 40°N and the longitudes 20°E and 23°E, 49 earthquakes with a magnitude ⩾5.3 and 28 earthquakes with a magnitude ⩾7.1 occurred. However, Galanopoulos (1967b, and personal communication to D.T. Snow, 1971) has pointed out that there had not been any significant earthquake in the Kremasta area during the period 1700—1965, whereas an area in the islands 100 km to the southwest has been very active. During the period 1951—1965, the epicenters were confined to the lower part of the Acheloos Valley, 40 km downstream of the dam. The Acheloos River area had no earthquake of magnitude >6 from 1821 until the commencement of the filling; however, three such earthquakes occurred between after the start of reservoir filling and 1970. Commencement of filling took place on July 21, 1965, and earthquakes started being felt in August. In an area of about 100 km × 100 km (shown in Fig. 51), 740 shocks occurred during August, 1965, to February 4, 1966. This was an unusually long series of shocks for Greece that more commonly experiences only two or three shocks in a month. There was a damaging earthquake of magnitude 6.2 on February 5, 1966, in an area that had previously not experienced earthquakes of magnitude greater than 5.5. The epicenter of this earthquake (39.1°N, 21.6°E) is very close to the northern

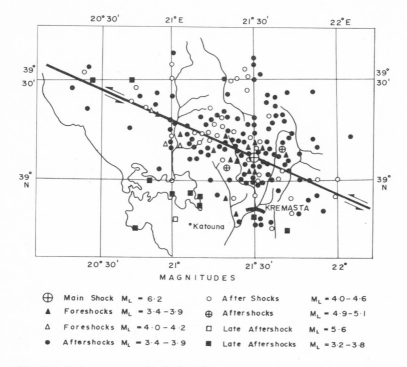

Fig. 51. Epicenters in the Lake Kremasta region, September, 1965, through November, 1966 (after Comninakis et al., 1968).

end of the lake, about 25 km north of the dam. This earthquake caused one death, injured 60 persons, created many landslides and slumps, and either the collapse or damage to 1,680 houses (*Seismological Bulletin of the National Observatory Athens*, February 1966). From February, 5, 1966, until the end of 1966, the number of shocks recorded was 2,580. The magnitudes of the recorded foreshocks and aftershocks reached to 5.6. Between August and December, 1965, four new stations were installed in the region of Greece with three components of short-period (T_s = 0.5 sec, T_g = 1.0 sec) Sprengnether-type seismographs having a magnification of 50,000. The nearest station at Valsamata on Kephallenia Island is about 115 km from the dam site. The epicenters were located by the four new stations of the network as well as by the central station in Athens. Accuracy of the epicentral determination has been mentioned to be 0.1° for earthquakes with $M_L \geqslant$ 3.4. The focal depth of the main shock was calculated to be about 20 km. Foci of the other shocks were estimated to be within the crust because of the existence of strong crustal phases. From macroseismic evidence, foci of small foreshocks were indicated to be very shallow and their epicenters were just below, or very close to, the lake (Comninakis et al., 1968). One may argue that the apparent increase of activity is due to the increased number of recording stations. But the swarms of felt earthquakes that started 45 days before the mainshock were unique. During January 15 to 20, 1966, six to eight tremors occurred daily. These had a peculiarly small felt area, a high intensity, a short duration, and a rumbling noise not experienced before. This indicates a shallow foci.

Reservoir level and earthquake frequency

The high rate of the water-level increase from November, 1965, to January, 1966, was immediately followed by an increase in the seismic activity

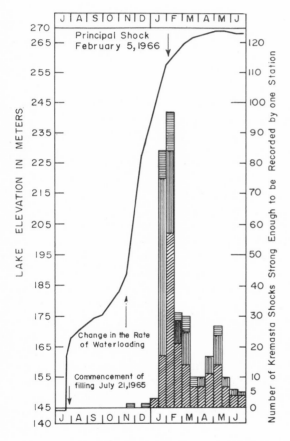

Fig. 52. Lake Kremasta elevations and earthquake frequency in the region (after Galanopoulos, 1967). Number of Kremasta shocks strong enough to be recorded by: ▰▰▰ = one station, ▥▥▥ = two stations, ▨▨▨ = several stations per 15-day period.

beginning in November, 1965, and a burst of tremors during January—February, 1966 (Fig. 52). The peak activity occurred in February, including the magnitude 6.2 earthquake of February 5, 1966, immediately following the long duration of an extremely high rate of loading. The activity decreased during March 1966, and afterwards when the reservoir level was more or less constant. It needs to be mentioned that tremors were initiated in the Kremasta area when the lake levels had risen to 245 m for the first time in January 1965; later the levels reached 268 m. Snow (1972) observes that no earthquakes now occur at these levels; however, during rainy seasons when levels exceed 270 m, tremors are felt in the region. The correlation coefficients computed between the reservoir level and tremor frequency are discussed later.

LAKE MEAD, UNITED STATES

Geology

The 142 m high Hoover Dam forms the reservoir called Lake Mead which has a $35,000 \times 10^6$ m^3 capacity and a maximum water depth of about 140 m. Lake Mead lies near the southeast margin of the Basin and Range Province and west of the Colorado Plateau. It begins to form where the Colorado River flows out of the Lower Granite Gorge of the Grand Canyon and then flows towards the west for about 80 km. The lake crosses several N—S-trending mountain ranges. The two basins containing the largest volume of water are the Virgin-Detrital Trough and the Boulder Basin, where the Colorado River turns southward and flows into Black Canyon (see Fig. 55). The Boulder Basin is just north of the dam, while the Virgin-Detrital Basin forms the central part of the reservoir. These basins are two distinct structural depressions separated by the elevated Black Mountains.

The geology of the Lake Mead area has been studied by many workers such as Longwell (1936, 1963), Longwell et al. (1965), Anderson (1971, 1973a), and Anderson et al. (1972). The salient geological features have been described recently by Rogers and Gallanthine (1974). We briefly review these features here.

As a majority of earthquakes have occurred in the Boulder Basin, it is of more interest and its geology is described in comparatively more detail. Fig. 53 shows the major tectonic and geologic features around this basin while Fig. 54 shows an E—W section through the area. Lake Mead was part of a wide sea during the Palaeozoic and Early Mesozoic era. Sediments, several thousands of meters thick, accumulated through the process of deposition and contemporaneous subsidence. Due west and north of Lake Mead, the thickness of sediments increases; about 8,000 m thick in the Spring Mountains west of Las Vegas. In the Mesozoic, large-scale uplift took place, this

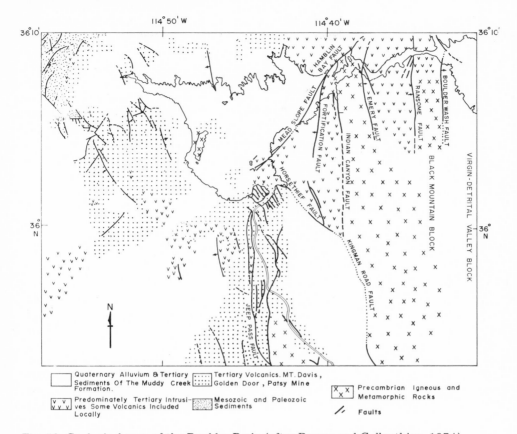

Fig. 53. Geological map of the Boulder Basin (after Rogers and Gallanthine, 1974).

was followed by mountain-building activity corresponding to the Laramide Orogeny. Subsequent erosion up to the Late Cretaceous has exposed the Precambrian surface near the Boulder Basin and south of it.

Large-scale volcanism started in the Early Tertiary. Five episodes of volcanism have been identified from the Late Cretaceous to Cenozoic. Most of the volcanics were deposited over the Precambrian basement. The thickness of volcanics varies considerably. However, thicker deposits are encountered due south of the Hoover Dam, where at certain places accumulations are of the order of 5,000 m. To the northeast of the Boulder Basin, volcanics have a thickness of about 900 m. The volcanics consist of basalts and andesitic flows associated with other volcanic products such as rhyolite glass, tuff, and breccia. Tertiary intrusive rocks are also present in a complex manner in the Precambrian rocks. In some areas intrusives lie conformably on the lavas as sills or laccoliths. Intrusives consist of quartz monzonites and granodiorites.

The present structural pattern of the Basin and Range Province developed

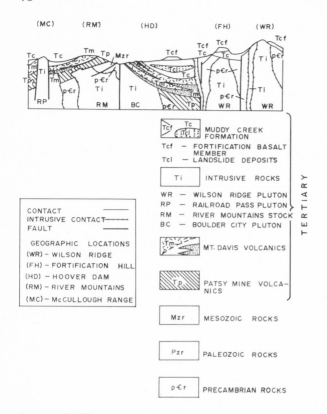

Fig. 54. E—W section through the Boulder Basin (after Anderson et al., 1972).

in the Miocene and Early Pliocene when volcanism accompanied by strike-slip and normal faulting continued. As basins with interior drainage were established, Muddy Creek sediments, which cover extensive areas of the Virgin-Detrital Trough and a region northwest of the Boulder Basin, were deposited. Deposits of the Muddy Creek Formation consist of conglomerates, sandstone, clay, basalt flows, salt, and gypsum. In the Virgin Basin area, salt and gypsum beds attain a thickness of a few thousand meters. A period of erosion, which developed the present course of the Colorado River, began in the Late Cenozoic. Since that time, the Colorado River has eroded much of the detrital fill and sediments from Boulder Basin, thus exposing many of the structures.

Faults are the main structural features surrounding the Boulder Basin. The Indian Canyon, Hamblin Bay, Boulder Wash, Horsethief, and the Jeep Pass are the important faults (Fig. 53). A major regional shear zone (the Las Vegas Fault) is known to pass north of Lake Mead. These structures, with the exception of the thrust faults in the north, developed during the Late

Tertiary together with the entire Basin and Range Province. The Hamblin Bay Fault is a NE-trending, left-lateral strike-slip fault with a displacement of about 19 km. Actually it is a part of a wide shear zone, 3—4 km wide at places, on which larger displacements have occurred. The Hamblin Bay Fault is believed to extend southwest beneath the Quaternary alluvium and Tertiary sediments in the Boulder Basin, where it branches out into the Mead Slope, Fortification, and Indian Canyon Faults.

The three latter faults and the Boulder Wash Fault are normal faults. Along the Boulder Wash Fault, either the Black Mountains block was uplifted with respect to the Virgin-Detrital block, or the Virgin-Detrital block subsided with respect to the Black Mountains block. Vertical displacements along these faults are of the order of 2,000 m. The connection between strike-slip faults and normal faults in the region is complex. According to Anderson (1973a), the strike-slip of the Hamblin Bay Fault was transformed into normal faulting at its south end along its branches at the conclusion of the tectonic activity.

Seismic activity is much greater in the Boulder Basin area although differential vertical displacement is much greater in the Virgin-Detrital Basin area. According to Anderson (1973b), the great contrast in the seismic activity of the two basins is probably related to the occurrence of widespread salt deposits at shallow depths in the Virgin-Detrital Basin area. The hydraulic continuity between the reservoir water and the deep hydrologic system beneath the Virgin-Detrital Basin is probably barred by the salt deposits and clay of low permeability. The Virgin-Detrital Basin area has a number of faults along which earthquakes could occur, releasing the deep subsurface stresses, but they were probably not reactivated in the absence of hydraulic continuity. Salt deposits have not been found to occur in the Boulder Basin area. Hydraulic continuity through permeable sediments and fractures to deep subsurface layers in the Boulder Basin area could be inferred from the vertical section drawn in Fig. 54.

Geodetic observations

The relative elevation changes, with respect to Cane Springs, that occurred around Lake Mead during the period from 1935 to 1963 are shown in Figs. 55 and 56 (Lara and Sanders, 1970 — adopted from Rogers and Gallanthine, 1974). The subsidence observed around Las Vegas is unrelated to the lake and is caused due to the withdrawal of groundwater, lowering of the water table and the consequent compaction of the sediments (Mindling, 1971). From the figures it is evident that the subsidence during the period from 1935 to 1949 dominates later elevation changes. On the basis of the elastic compression of an infinite half space, Westergaard and Adkins (1934) had predicted the settlement of the lake area and had also theoretically computed the displacement due to isostatic compensation of a crustal block. A

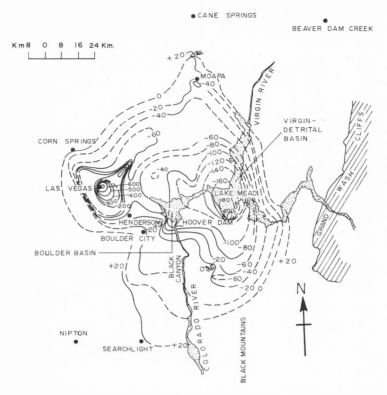

Fig. 55. Contours of elevation changes (in mm) in the interval 1935 to 1963—1964 relative to Cane Springs. Solid lines correspond to the Survey of 1963 and dotted lines to that of 1964 (after Lara and Sanders, 1970).

comparison of these predictions with the precise level measurement results (Raphael, 1954) showed that the elastic compression model was adequate for the explanation of the observations, whereas isostatic compensation would require three to four times the displacements observed until 1949.

Rogers and Gallanthine (1974) have computed the displacement expected for the seven earthquakes of magnitude 5.0 and one of magnitude 4.8 which occurred during 1939—1963. This totals to a displacement of 27.5 cm, and is 40% larger than the maximum observed subsidence at Lake Mead during the same period.

Seismicity

The seismicity associated with Lake Mead has been frequently cited as a classic example of earthquakes caused by reservoir loading. Filling of the lake began in 1935. Before the filling commenced, there was no historical

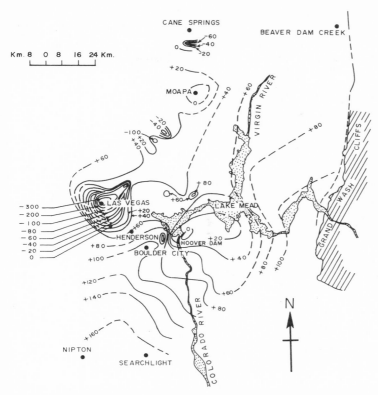

Fig. 56. Contours showing elevation changes in the interval 1949 to 1963—1964 (after Lara and Sanders, 1970). For details see Fig. 55.

record of earthquakes and the area was considered aseismic (Jones, 1944; Raphael, 1954; Simon, 1972). During 1937, when the lake reached the peak load of the year, about 100 tremors were felt locally (Carder, 1945). This activity continued, reaching a maximum in 1939, culminating in a magnitude 5 earthquake. The seismic activity has continued until recently with numerous felt earthquakes. Microearthquakes of local magnitudes less than 1.0 occur even now at a rate of one or two per day.

Seismic surveillance started in 1937 when 3 strong-motion seismographs were installed at and near the dam. In 1938, a Wood-Anderson seismograph was installed at Boulder City. A tripartite array, about 80 km on each side, began operating around the lake in 1940 with vibration meters. These instruments were replaced in 1942 by short-period three-component Benioff seismographs which operated for 10 years. A Benioff seismograph has also been operating at Boulder City since 1952. The net of stations established by N.O.A.A. in 1960 to monitor the seismicity near the Nevada test site encompasses most of the lake area. Microtremors at Lake Mead are being recorded

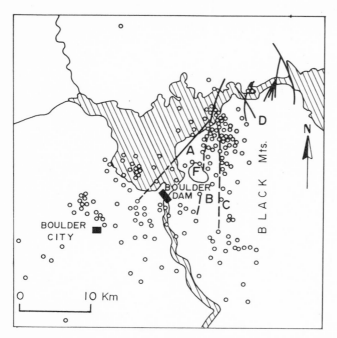

Fig. 57. Epicenter locations in the Lake Mead area from June, 1942, to December, 1944. A, B, C and D are the faults as shown in Fig. 53, F is the Fortification Hill (after Carder, 1945).

by an array of 10 stations which was established on July 6, 1972, around the Boulder Basin; the extent of the array is 20 km.

Most of the epicenters located by Carder (1945) and Rogers and Gallanthine (1974) are within 25 km of the lake. For the period from 1936 to 1944, the epicenters concentrated along the existing faults (this is clearly seen from the epicenters plotted for the period from June, 1942, to December, 1944, in Fig. 57) with focal depths of less than 9 km. Error in epicentral location was estimated to be 1 km or less (Carder, 1945). One-half of the events located by Rogers and Gallanthine (1974) have standard errors of 0.5 km or less. Some events, that lie outside the array to the south and southwest, have the largest errors, i.e. about 2.0 km. The events for a period of one year beginning from July 6, 1972, are shown in Fig. 58. The more accurately determined epicenters among these are related with the fault system (Rogers and Gallanthine, 1974). The strongest trend of epicenters begins from near the intersection of the Hamblin Bay, Mead Slope and the Fortification Faults. It extends further southwestward, parallel to the Mead Slope Fault, terminating about 5 km south of the lake shore. A similar trend may exist beneath the lake. The deepest cluster of foci, which have a depth of 5—6 km, occurs east of the Mead Slope Fault. Near the center of the lake,

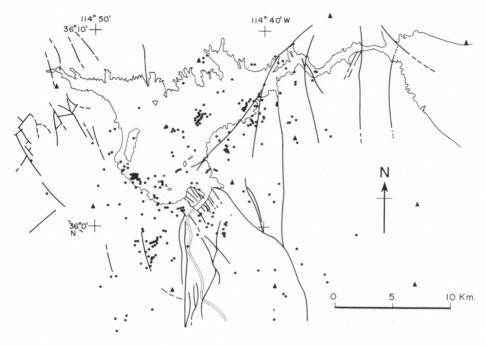

Fig. 58. Epicenter locations in the Lake Mead area from July, 1972, to June, 1973. Seismograph station locations are shown by triangles. Major faults are superimposed (after Rogers and Gallanthine, 1974).

clustering takes place at 4—5 km depths, while the depths vary from 1½ to 2 km for the cluster near the southwest shore.

Reservoir level and earthquakes frequency

Carder (1968, 1970) and Mickey (1973a) found that there was a positive correlation between the seasonal lake load and seismic activity during a number of years after the lake was filled. However, during later years a negative or zero correlation has been observed. Earthquakes in the area were felt for the first time in September, 1936, when the lake level reached its maximum of that year. Fig. 59 shows the variations in Lake Mead water levels and the frequency of the tremors. The maximum seismic activity and the strongest earthquake, which had a magnitude of 5, occurred in May, 1939, when the lake level again rose after attaining the normal level. The two other significant uprises in water level are in 1941 and 1942; these are both followed by significant seismic activity within a few weeks of the peak levels. The other bursts of seismic activity also seem to follow uprises in water level.

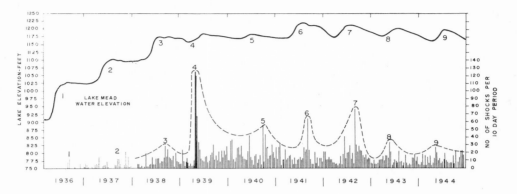

Fig. 59. Lake Mead water levels and the local seismicity. For 1936 and 1937, only the felt shocks are plotted. The uprises in water levels and the corresponding bursts of seismic activity are numbered. General trend of tremor-frequency variation is shown by dotted lines (after Carder, 1945).

Statistical analysis of water level and earthquake frequency at Lake Mead will be discussed later.

DENVER, UNITED STATES

A deep well was drilled at the Rocky Mountain Arsenal northeast of Denver, Colorado, to dispose of contaminated waste water. The depth of the well is 3,671 m. Injection of fluid began on March 8, 1962, and continued until September 30, 1963, at an average rate of about 21 million liters per month. Fluid was not injected from October, 1963, to August, 1964. Then fluid injection was resumed under gravity flow at an average rate of about 7.5 million liters per month until April 6, 1965. This was followed by injection under pressure at an average rate of 17 million liters per month. Within a few weeks of the beginning of fluid injection, a swarm of tremors, including some strong earthquakes, with epicenters near the well started in April, 1962. Injection was terminated on February 20, 1966, because of a suggested causal relationship between fluid injection and the Denver earthquakes (Evans, 1966).

Colorado is considered to be a region of minor seismicity. From newspaper records dating back 100 years, Hadsell (1968) listed about 30 felt earthquakes in Colorado, prior to the construction of the well. Simon (1969) reported that instrumentally recorded earthquakes from 1966 to 1968 occurred in the same regions where earthquakes were felt earlier. However, most of the historical earthquakes have occurred west of the Rocky Mountains while Denver is on the east side. Prior to the start of fluid injection in 1962, there is knowledge of only one earthquake in the Denver area which

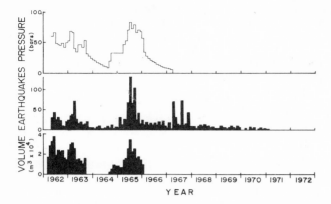

Fig. 60. Monthly number of earthquakes in the Denver area, monthly volume of injected water and wellhead pressure in the disposal well (after Handin and Raleigh, 1972).

occurred in 1882 and had an intensity of VII (MM scale). The data collected for this historical earthquake could not be held to exclude the possibility of a large number of earthquakes being triggered by fluid injection.

Evans (1966) pointed out a distinct correlation between the monthly tremor frequency and the monthly amount of injected water for the period 1962—1965 (Fig. 60). Injection ceased in early 1966, and the frequency of earthquakes diminished. The frequency increased, however, in 1967, more than a year after termination of fluid injection, when three earthquakes with magnitudes of 5—5.2 occurred which also caused some damage. The larger

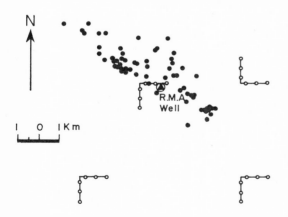

• Earthquake Epicenter.   ○─○─○ Recording Location

Fig. 61. Epicenters of earthquakes near Denver deep-disposal well located in January and February, 1966, by means of a dense network of seismic stations (after Healy et al., 1968).

shocks continued even up to 3 years after cessation of the fluid injection. In 1969, two shocks of magnitude 3.5, and fourteen more of magnitude $\geqslant 2.5$ occurred, which were felt in the area. The stronger earthquakes, which occurred at a distance of 5 km from the well after the termination of fluid injection, were difficult to explain in terms of fluid injection. However, Healy et al. (1968) have tried to explain them as due to an advancing pressure front, as discussed in Chapter 6. The earthquake locations were not accurately known until 1966, four years after injection began. Subsequently, with the help of eight L-shaped arrays, each having 6 seismometers at 1/2-km intervals in a small area around the well, sixty-two earthquake hypocenters were very accurately located in a period of 2 months. These epicenters were found to lie along a linear trend 8 km long by 2 km wide passing through the disposal well, with focal depths ranging between 4.5 and 5.5 km (Fig. 61). The focal depths correspond closely to the injection depth of 4 km. The locations are correct to 1 km (Healy et al., 1968). The total number of earthquakes exceeded 1,500 during the period 1962—1967. Many were in the magnitude 3—4 range and were felt over wide areas. The triggering mechanism of the Denver earthquakes by fluid injection is discussed in Chapter 6.

RANGELY, UNITED STATES

Oil is produced in the Rangely oil field from a closed anticline in the Uinta Basin in Colorado. Weber sandstone, a Late Palaeozoic quartz sandstone at a depth of 1,830 m, is the producing horizon. From the initiation of production in 1945, the formation fluid pressure decreased rapidly until 1957, when water was injected for secondary recovery.

In November 1962, the Uinta Basin seismological observatory was installed at 65 km west-northwest of Rangely. This station immediately recorded a large number of small earthquakes in the vicinity of Rangely (Raleigh, 1972), and until January 1970 it recorded 976 earthquakes from the oil field, of which 320 earthquakes were of a magnitude larger than 1 (Gibbs et al., 1973). Changes in the annual number of earthquakes seem to correlate with the volume of fluid injected per year. In 1967, a local network of portable stations (Pakiser et al., 1969) showed that these earthquakes clustered at two places, i.e. in the northwest end of the field and along the south-central margin of the field (Fig. 62). At both these places, the pore pressure due to water flooding exceeded the original field pressure.

Accurate locations of the earthquakes became available in 1969 when a local network of 14 permanent seismograph stations was installed. Between October, 1969, and November, 1970, about 1,000 earthquakes were recorded which were mostly from the south-central part of the field. The magnitudes of these earthquakes range between $M_L = -0.5$ and 3.5. These epicenters lie along the southwest extension of a fault mapped at the top of the

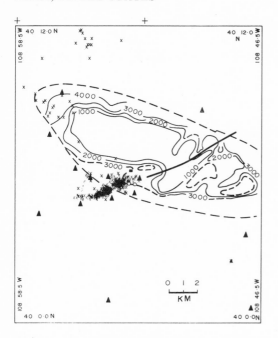

Fig. 62. Epicenters of earthquakes in the Rangely oil field (outlined by the dashed line) between October, 1969, and November, 1970. Contours give the bottom-hole pressure (in psi) measured in September, 1969. Triangles represent seismograph stations; circles, experimental wells; square, well in which stresses were measured by means of hydraulic fracturing (after Raleigh, 1972).

Weber sandstone from well-log data. Here, the pore pressure was greater than the original field pressure. These earthquakes occurred at depths ranging between 1,830 and 3,660 m within or below the injection zone.

DALE, UNITED STATES

In addition to the earlier-mentioned two well-known examples of tremors originating following fluid injection at Denver and Rangely, Sykes et al. (1973) have drawn attention to similar occurrence at Dale, New York. The Lamont-Doherty Geological Observatory has installed a network of 18 seismograph stations in New York State and adjoining areas to study the distribution of small earthquakes. Eight of these stations were installed in Western New York to monitor any changes in seismic activity that may be associated with the operation of deep disposal wells in the Buffalo area. A high-pressure injection well for hydraulic salt mining was connected to a series of recovery wells by hydrofracturing at Dale in July, 1971. These wells bottomed in or very close to the Clarendon—Linden Fault zone. Before the injection, the seismic activity close to the region was restricted to about one event every few months. Following the injection, a station about 1.5 km away from the wells detected about 100 clearly observed events every day. In November, 1971, the high-pressure injection was shut down and the seismic activity decreased within two days to approximately the pre-injection level. None of the events had a magnitude larger than 1.5; however, some of them were felt locally. The better located events, which were recorded with the help of the portable seismographs, fell within 5 km of an elongated zone sub-parallel to the strike of the Clarendon—Linden Fault.

A similar experiment was repeated in August, 1972. Another injection well, 0.3 km north of that drilled in 1971, was connected by hydrofracturing to the adjacent recovery wells. Although the pressure history of this well was similar to that of 1971, only a few seismic events were observed by a much more dense seismic network. Hydrofracturing in 1971 was performed near the basal contact of the salt with the highly competent Lockport dolomite, whereas in 1972 it was confined to the center of the salt layer. It seems that the fluid injected in 1971 gained access to the fault zone and triggered the release of tectonic strain, while in 1972 the fluid was confined to the salt only. This hypothesis is supported by the fact that the water loss was negligible during 1972 compared to the higher loss in 1971, when earthquakes were generated.

TALBINGO DAM, AUSTRALIA

The Talbingo Dam is a part of the Snowy Mountain Hydroelectric Scheme in southeastern Australia. It is a 162 m high earth- and rock-filled structure with a crest length of about 700 m. The capacity of the Talbingo reservoir is 935×10^6 m^3.

A local network of four seismograph stations has been operating in the area since 1957. As Talbingo falls outside the quadrangle of this network, an additional station was established about 3 km north of the Talbingo Dam in 1969, two years before the filling of the Talbingo reservoir, in order to carefully monitor the seismic events related to the filling of the reservoir (Fig. 63). Three portable stations were later installed near the reservoir area in July 1971.

Although the Snowy Mountains region is an area of minor seismicity, in the 13 years of monitoring by the local network, including the 2-year period of close monitoring from the Talbingo station, only one minor event occurred within 25 km of Talbingo before the reservoir filling. Soon after reservoir filling commenced in May, 1971, activity was observed on the records of the dam-site station. Reservoir filling began on May 1, 1971, and the tremors commenced on May 19. Three tremors occurred in the month of May, in June the activity increased to 39 events. In July and August, a

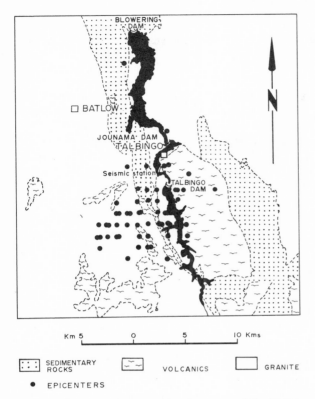

Km 5   0   5   10 Kms

⬚ SEDIMENTARY ROCKS   VOLCANICS   GRANITE

● EPICENTERS

Fig. 63. Geological map and epicenters of the Talbingo reservoir area. The northernmost epicenter is that of the pre-impounding tremor on May 1, 1971 (adapted from Timmel and Simpson, 1973).

further increase in the seismic activity, corresponding with the increasing water-level, was noted. After August, there was a decrease in tremor frequency, corresponding to a sharp drop in the rate of filling. The reservoir had risen to within a few feet of its maximum level during December, 1971, and remained fairly constant during the following years. The seismic activity had a general low level in early 1972, but later it again increased (Timmel and Simpson, 1973). Fig. 64 suggests a relationship between the rate of filling and seismicity.

In the first 15 months following the commencement of filling, over 3,000 microearthquakes were detected. Among these, 100 events were large enough to be located by the regional network. The largest of these events had a magnitude of 3.5 and about 60 events had magnitudes greater than 1.5. In spite of the large number of events recorded during filling, the seismic activity is small in terms of the total energy released. The epicenters have been found to be very shallow and are confined to the immediate vicinity of

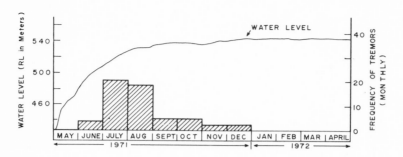

Fig. 64. Water level variation of the Talbingo reservoir and monthly tremor frequency in the area (after Timmel and Simpson, 1973).

the reservoir (Muirhead et al., 1973). Their preliminary locations, determined without using the data of the temporary stations near the lake, are shown in Fig. 63. The epicenters are within 10 km of the reservoir. The epicenters for the year 1972, which are not shown in the figure, display a shift towards the eastern side of the reservoir (Timmel and Simpson, 1973).

HENDRIK VERWOERD DAM, SOUTH AFRICA

The Hendrik Verwoerd Dam (30°38′S, 25°30′E) in South Africa is situated on the Orange River and forms the first phase of an extensive development scheme on this river. The 66 m high and 600 m long, double-arch, concrete wall dam creates a reservoir of 5,000 × 10⁶ m³ capacity. Filling of the reservoir commenced in September, 1970.

Green (1974) has described the geology of the region. The dam lies in a relatively stable area, near the center of a large basin of gently disposed Upper Palaeozoic and Lower Mesozoic sediments of the Karoo System. At the dam site, the horizontal Beaufort Series of Palaeozoic age consists of mudstones, sandstones, and shales. These have been extensively intruded by dolerite dykes and sheets. Except for the effects of these intrusions, the sedimentary beds are relatively undisturbed and there are no major faults in the area. However, minor faulting is evidenced. Numerous intrusions govern to a much large extent the local topography.

A single-component seismograph was operating intermittently near the dam from 1966 to 1970. Two months before the commencement of filling, a three-component system was installed at the same place. Three months after the dam closure, four additional stations were installed having a 36-km maximum baseline of the network around the dam (Fig. 65). An array of tiltmeters was also installed at,the same time. One seismograph very close to the dam began operating during May, 1971.

In the past, four earthquakes with an estimated magnitude exceeding 5.0

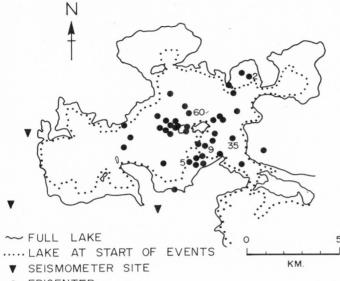

FULL LAKE
LAKE AT START OF EVENTS
SEISMOMETER SITE
EPICENTER

Fig. 65. Hendrik Verwoerd reservoir and the location of epicenters (after Green, 1974).

have occurred within 100 km of the dam. The largest earthquake, which had a magnitude of 6.0, occurred in 1912. The most recent event had a magnitude of 5.1, and it occurred in 1955. However, no local shocks were recorded by the single seismograph operating from 1966 to 1970. The regional network, which was established later, also did not show any activity for about six months following the impoundment. The activity began on Febru-

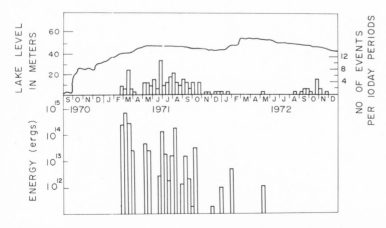

Fig. 66. Reservoir level, tremor frequency, and energy (after Green, 1974).

ary 27, 1971, six months after impounding, when the lake level was 40 m high. Ninety-seven local earthquakes were recorded in the succeeding 10-month period (Fig. 65). The level of seismicity declined after December, 1971, and there was no marked increase in the following rainy season when only 25 small events of magnitude $\geqslant 2.0$ were recorded (Fig. 66).

Green (1974) computed the theoretical crustal depression beneath the lake using the method of Gough and Gough (1970a). The maximum water pressure is 6.3 bars giving a maximum depression of 31.7 mm at 1 km depth. The events are concentrated within 3 km of the point of maximum depression and have a maximum focal depth of 6 km.

VAJONT DAM, ITALY

The Vajont Dam is in a deep gorge and is at present the highest dam in the world. The height of the dam is 266 m and it provides a storage of 150×10^6 m³. The dam is founded on Middle Jurassic limestone. The area has massive dolomitized Upper Jurassic marls, clays, and limestones underlying Cretaceous formations. The top strata of the Jurassic is known as Malm, which is a calcareous marl. The geology of the area is described in Fig. 67 and two sections are given in Fig. 68.

On October 9, 1963, the world's most tragic dam disaster occurred when an enormous mass of Cretaceous strata, with an estimated volume of 250—300×10^6 m³, broke off in one piece from Mount Toc on the left side of

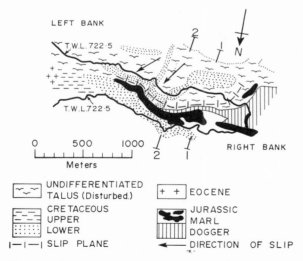

Fig. 67. Geological map of the Vajont reservoir area. The dam is on the right, the top water level is 722.5 m O.D. The slip plane and slip direction of the land slide are indicated (adapted from Walters, 1971).

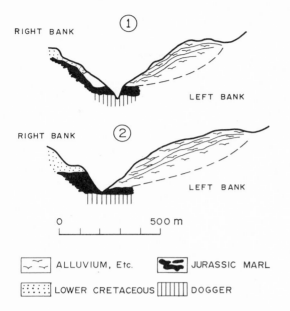

RIGHT BANK

LEFT BANK

RIGHT BANK

LEFT BANK

0   500 m

☐ ALLUVIUM, Etc.   ■ JURASSIC MARL

☐ LOWER CRETACEOUS ☐ DOGGER

Fig. 68. Vajont reservoir. Suggested slip sections *1* and *2* (see Fig. 67) of the slide (after Walters, 1971).

the valley and fell into the reservoir. The volume of the overflow, $25-30 \times 10^6$ m^3, created a huge, 70 m high wave. This wave erased the town of Longarone and claimed about 2,000 lives.

Reservoir filling commenced in 1960 and a seismograph station was established at the same time. About 250 tremors were recorded near the Vajont Dam over the period 1960—1963. The filling of the lake produced these weak tremors, and their recurrence disturbed the equilibrium of the strata (Caloi, 1966). Epicenters have been located within 3—4 km of the dam with the aid of a seismic station which was operated at the dam site.

The water level in the Vajont reservoir shows a good correlation with the tremor frequency. The three significant uprises in the Vajont reservoir are followed by three bursts of seismic activity, as clearly shown in Fig. 69. Following every peak, the decrease in water level is followed by decreased seismic activity. The earthquake in May and later the burst of earthquakes during October—December, 1960, correspond well with the seismic activity throughout the year of 1962, when about 110 local tremors occurred. This is followed by a quiescent period from January to April, 1963, perhaps because of the decrease in the water level from December, 1962, to March, 1963. During the third uprise, the rate of loading had been quite high and the highest level was reached in September, 1963. Conspicuous seismic activity was observed from May to September, 1963. The maximum activity of September, 1963, was followed by the disastrous landslide of October 9,

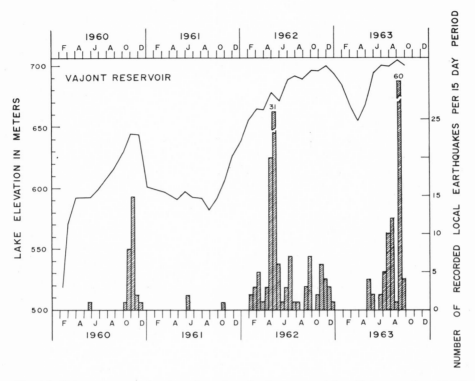

Fig. 69. Vajont reservoir levels and earthquake frequency (after Galanopoulos, 1967).

1963. Later, during 1964, 1965, and 1966, each rise of water level corresponds well with an intensification of seismic activity. Due to a flood in September, 1965, about 400 shocks were recorded within a few days (Caloi, 1970). When the water level decreased, activity also disappeared. It again increased many-fold during the floods in November, 1966, when the water level rose abruptly by 60 m.

MONTEYNARD DAM, FRANCE

The Monteynard Dam is located in a narrow valley of the Drac River in the French Alps. It is an arch dam of 130 m height with a 210 m long crest. The ground consists of Lias limestone which dips nearly 45° parallel to the valley. The reservoir has a normal storage of 275×10^6 m^3 (Rothé, 1969).

Filling commenced in April 1962, and the dam was filled to capacity by April 15, 1963. The tremors began after a few days. A damaging surface-focus earthquake of magnitude 4.9, which was preceded by two weak tremors, occurred on April 25, 1963, close to the reservoir. There has been a

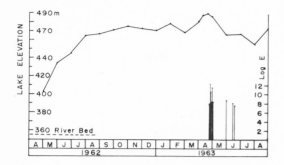

Fig. 70. Monteynard. Lake elevation and seismic energy released (in joules) (after Rothé, 1970).

subsequent occurrence of mild tremors. A swarm of shocks was recorded during 1966 from the same region, the largest shock being of magnitude 4.3. The seismograph station at Roseland, situated at a distance of 110 km from the dam, and the Monteynard station which was established on August 9, 1963, recorded from the area 15 tremors in 1963, 9 in 1964, 1 in 1965, 23 in 1966 and 16 in 1967.

Rothé (1968) showed that a close relationship exists between the reservoir level and tremor frequency or the energy released in the tremors. No earthquake was recorded from the region of Monteynard itself before the dam was built, although in April, 1962, a strong tremor shook the Vercors area, about 12 km northwest of Monteynard. The tremors started a few days after the reservoir had reached its maximum depth of 135 m (Fig. 70). The two largest earthquakes of April 25, 1963, and August 24, 1966, occurred when the level was at its highest.

The dam is located in a tectonically disturbed area. The Drac Gorge contains a number of faults, fissures and diaclases which are parallel to the gorge and might have caused considerable water seepage.

GRANDVAL DAM, FRANCE

The Grandval Dam is located on the Truyere River in the Hercynian Mountains called the Central Massif. The area had a slight seismic activity. The reservoir capacity is 292×10^6 m^3 with a maximum water depth of 78 m.

Filling commenced on September 15, 1959. After the first filling was completed in March 1960, and the lake was full, some earthquakes, accompanied by underground noises, were felt by the workers of the hydroelectric plant, particularly on December 31, 1961, and January 1, 13, and 14, 1962. The lake was filled again and during August—September, 1963, the maximum water depth of 78 m was reached. A strong shock was felt on August 5, which had an intensity of V (MM scale). Another strong shock occurred 2 months later. These two shocks were the strongest and occurred after the rapid rate of loading. These two shocks in 1963, and the two other shocks in 1964 with an intensity up to V, whose foci were located exactly under the reservoir with very shallow depths by 11 French seismograph stations, were felt at the dam and nearby villages (Rothé, 1969).

LAKE MARATHON, GREECE

Lake Marathon is a relatively small lake situated partly over the Tertiary sediments and partly over the crystalline schist and granite in the Attica Basin, Greece (Fig. 71). Filling began in 1929 and earthquakes were first felt in 1931 when the lake attained its peak level for the first time (Galano-

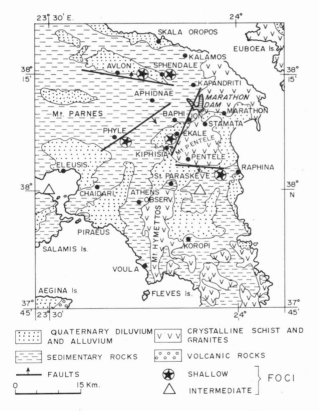

Fig. 71. Marathon Dam. Geological map of the area and the earthquakes which occurred since the filling of Lake Marathon in June, 1931 (adapted from Galanopoulos, 1967).

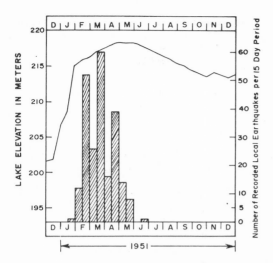

Fig. 72. Lake Marathon. Lake elevation and local earthquakes ($3 \geqslant M_L \geqslant 1$) recorded at Athens during 1951 (after Galanopoulos, 1967).

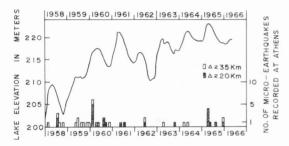

Fig. 73. Lake Marathon. Seasonal fluctuations of the water level and microearthquakes occurring within 35 km of the lake for the period 1958–1965 (after Galanopoulos, 1967).

poulos, 1967). Two damaging earthquakes of magnitude >5 occurred in 1938. Galanopoulos has plotted the earthquakes' epicenters around Lake Marathon (Fig. 71). The epicenters fall within 15 km of the lake.

The swarm of earthquakes during 1951 correlates very well with the rise in water level (Fig. 72). The seismic activity increased following the loading in January, 1951, and seismic activity maxima occurred just after a long duration of rapid loading. Activity diminished as the water level decreased. Galanopoulos (1967) has shown that the seasonal fluctuations in the lake levels from 1958 to 1966 were followed by quite marked seismic activity and that the occurrence of earthquakes is proportional to the rate of change of reservoir level (Fig. 73). He has also shown that most of the strong earthquakes from 1931 to 1966 occurred during periods of a rapid rise in water level.

NOUREK DAM, U.S.S.R.

N.I. Nikolaev (personal communication, 1973) has brought to our attention the case of induced seismicity at the Nourek Dam which is constructed in the middle course of the Vakhsh River in Tadjikistan, U.S.S.R. It would be the highest embankment dam in the world having a project height of 300 m. By 1973, the dam had reached one-half the project height. Filling started in late 1972, a maximum water head of 140 m was attained in the first stage. The dam's engineers are filling the reservoir at a slow rate to avoid the possible triggering of strong earthquakes. The full reservoir would be 75 km long and 1 km wide with a capacity of $11,000 \times 10^6$ m^3 and a maximum water head of 250 m.

The Nourek Dam is situated in the Poulissang Gorge within the Tadjik Basin where intense Neogene/Quaternary (neotectonic) movements occur. The region has a complicated geological structure. Nikolaev has distinguished three stages in the geological history of the Tadjik Basin: (1) Hercynian geosyncline regime, (2) Mesozoic/Cenozoic (Paleogene) platform regime, and (3) Neogene/Quaternary (neotectonic) orogenic regime. The two upper stages are exposed in the region, and consist of marine and platform sediments and orogenic continental molasses. A gypsum-salt series also occurs with specific forms of salt tectonics. Irregular tectonic movements took place during recent times. The Vakhsh River valley in the Poulissang Gorge was widened due to such movements.

Before the construction of the Nourek Dam, intense seismicity with earthquakes of intensities in the range of VIII to IX (MM scale) was recorded in the region. During the first stage of filling, water was raised to the 100-m level by December, 1972. The seismic activity, which was recorded by seismographs in the reservoir area and neighboring sites, strongly increased in the course of filling. Certain epicenters were located beneath the reservoir. The number of weak tremors increased ten times within an area of 5 km from the reservoir. Before the reservoir filling, the number of tremors was an average 3—4 per 10-day period. The frequency increased to 30—40 after filling of the reservoir. The focal depths of these shocks are within 5 km. Only one such earthquake with a magnitude of 4.5 occurred in the 2-year period of recording before the filling while the earthquakes which are now observed are more frequently in the magnitude range of 4—4.5. All these earthquakes have occurred within 15 km of the reservoir.

KUROBE DAM, JAPAN

The Kurobe Dam (36°34′N, 137°40′E) is an arch dam of 180 m height, constructed over biotite-granite bed rock in central Japan. Capacity of the reservoir is 149×10^6 m³, with a maximum water depth of 180 m. Hagiwara and Ohtake (1972) have demonstrated that local earthquakes near the Kurobe Dam are related to the filling of the reservoir. Fig. 74 shows the shallow earthquakes near the Kurobe Dam during the 45-year period between 1926 and 1970. Small and moderate earthquakes have frequently occurred in this mountainous region. The seismic area of the Matsushiro earthquakes lies 50 km east of the dam; however, the Kurobe Dam area was free from notable shocks during the 35 years preceding the start of filling in 1960. All the three significant shocks within 10 km of the dam occurred after the construction of the dam, i.e. on August 19 (M_L = 4.9) and 21 (M_L = 4.0), 1961, and November 16, 1968 (M_L = 3.8). These shocks were located to be near Mt. Harinoki on the eastern shore of Lake Kurobe. The last shock took place about 15 days after the water level rose beyond its normal level of 1,443 m (depth = 175 m) towards the end of October, 1968. The first two shocks also seem to be related to the filling of the lake, when the maximum water

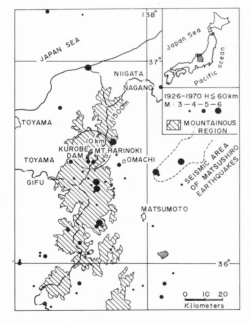

Fig. 74. Kurobe Dam. Map showing shallow earthquakes with focal depth ≤60 km in the mountainous region of central Japan for the period 1926—1970. Matsushiro earthquakes are not individually plotted on the map (after Hagiwara and Ohtake, 1972).

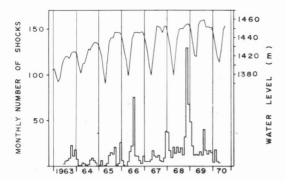

Fig. 75. Kurobe Dam. Fluctuation of the water level of the reservoir and monthly frequency of dam-site shocks (after Hagiwara and Ohtake, 1972).

depth was 102 m. Furthermore, many microearthquakes occurred around the dam during the impoundment of the reservoir.

Hagiwara and Ohtake analyzed the records of the seismograph station at the Kurobe Dam site (36°34′N, 137°40′E) for the 84-month period from May, 1963, to April, 1970. The seismographs are three-component ones of "HES 1—0.2" type with a magnification of 50,000. This period of analysis includes the active stage of the Matsushiro swarm earthquakes. Earthquakes belonging to this swarm registered an S minus P interval varying from 4 to 8 seconds at the Kurobe Dam observatory. It has therefore been possible to exclude them by considering only the very local shocks with an S minus P interval of 1.0 second. The number of such dam-site shocks was 1,182 within the 84-month period of investigation. Fig. 75 shows the monthly number of shocks compared to the water level of the reservoir. Filling began in 1960 and the peak level was reached in July, 1969, with a depth of 180 m. Comparing the two curves it can be seen that a notable increase in the seismic activity occurs when the reservoir has a higher water level, for example, in July 1966, December 1967, and November 1968. With lower water levels, the seismic activity shows a tendency to diminish. The correlation coefficient and the time lag between water level and tremor frequency obtained at the Kurobe Dam are discussed later.

OUED FODDA DAM, ALGERIA

Oued Fodda in Algeria provided one of the earliest examples of earthquakes associated with reservoir filling (Rothé, 1969). The dam is 89 m high with a crest length of 170 m. The reservoir's capacity is 225×10^6 m³. Gourinard (1952) has related the earthquakes near the Oued Fodda Dam with the reservoir. Filling of the reservoir commenced during the end of

1932. Earthquakes were frequently felt near the dam from January to May, 1933. No shock has occurred since then. The geology of the area had been studied in detail for the selection of the dam site. The seismic risk of the region was evaluated and the permeability of rocks was studied. The decision to build a gravity dam rather than an arch dam was made on the ground that limestones were considered inherently too weak to take the thrust of an arch dam. The dam is founded on Jurassic limestones which dip steeply upstream. Downstream of the dam, the Triassic limestones are full of grottos and springs, they are dolomitized and frequently develope karstic features. Alternating hard and soft strata of marls and limestones of Neoconian and Upper Jurassic age occur in the reservoir area.

CONTRA DAM, SWITZERLAND

The Vogorno reservoir is situated behind the Contra Dam in Switzerland. Filling of the lake commenced in August, 1964, and seismic activity began in May, 1965, when underground noises and several hundred local tremors occurred (Susstrunk, 1968). According to Lombardi (1967) these tremors followed a rapid rise of water level. The epicenters of these earthquakes were localized in a zone bounded by two faults in Berzona, close to the reservoir. According to Susstrunk (1968), the variations in lake level were followed by a similar change in the frequency of shocks which occurred with a time lag of 3 to 6 weeks. The strongest shocks occurred in October and November, 1965, after a few weeks of maximum lake level. Though the extent of damage is not known, the shocks were strong enough that evacuation of the village of Berzona was considered necessary at one stage. The lake was subsequently emptied and refilled; however, the activity did not increase and after some time it ceased. According to Lombardi (1967) a new equilibrium was later reached which was not disturbed by successive fluctuations of the water level.

MANGLA, PAKISTAN

The Mangla reservoir is situated 200 km north-northwest of Lahore, Pakistan, where the Jhelum River leaves the high country of Kashmir and enters the flat plain of the Indus Basin. The Mangla reservoir is enclosed by three dams, the Mangla, Sukian and Jari (Fig. 76). It has a capacity of $7,250 \times 10^6$ m^3, it comprises an area of 250 km^2, and has a mean depth of 26 m (Adams, 1969). The maximum height of the dam is 135 m. The region lies in the seismically active Himalayan belt. A short-period Willmore seismograph was installed at Mirpur in late 1965 and two more stations at Baral and Jari were installed when impounding began in February, 1967. Adams and Ahmed

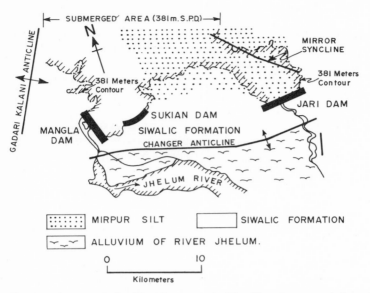

Fig. 76. A part of the Mangla reservoir. The Mangla reservoir is enclosed by three dams —
Mangla, Sukian and Jari. Perimeter of the reservoir follows the level of 1250-ft (381 m)
Survey of Pakistan Datum (S.P.D.) (adapted from Walters, 1971).

(1969) mention that the overall occurrence of the detected earthquakes
increased by a factor of 2.5 after impounding. However, this increase could
mostly be accounted for as a result of the addition of new stations and any
connection between the filling of the reservoir and the increase in seismicity
remains unproven (Adams, 1969). Most of the earthquakes had small magnitudes being of the order of 1—2 and were widely distributed (Fig. 77). The
largest earthquake had a magnitude of 3.6. Fig. 78 shows the frequency of
earthquake occurrence and seismic energy release in relation to the water
levels of Mangla reservoir. The average monthly energy release increased from
3.0×10^{15} ergs before impounding to 5.6×10^{15} ergs subsequent to impounding. Since these figures are largely controlled by the few shocks of
largest magnitude, the installation of new stations is not likely to influence
them. However, Adams and Ahmed (1969) felt that the increase in the
energy release is small and of doubtful significance. They pointed out that
there is some relationship between the changes in the level of the reservoir
and the local seismicity. Fig. 78 shows that the number of tremors increased
during the period March—December, 1967, following the initial impounding.
Another increase in the tremor frequency is noticeable during July—August,
1968, at the time when the reservoir was rapidly loaded. During the period
April 1 to July 12, 1968, when the reservoir was at its lowest level, no
shocks were recorded.

In the area, the basement rock is overlain by an estimated 3 km of friable

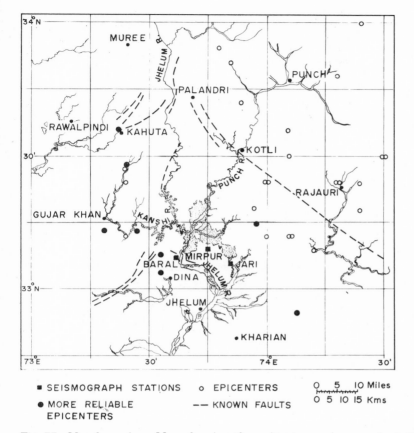

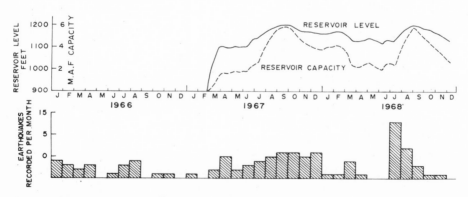

Fig. 77. Mangla region. Map showing the epicenters, seismograph stations and known faults (after Adams and Ahmed, 1969).

Fig. 78. Mangla reservoir. Frequency of tremors and reservoir water level (after Adams and Ahmed, 1969).

sandstones, siltstones, and clays of the Siwalik (Late Tertiary) Series which runs parallel to the Himalaya. These rocks offer little resistance to any applied load, and are expected to gradually deform to accommodate it. According to Adams and Ahmed (1969), the seismic effect of the Mangla reservoir is not pronounced due to this effect. The Siwalik Formations were uplifted in Early Pleistocene time and we find them as anticlines, such as the Changer Anticline which runs E—W, 1—3 km south and downstream of the Mangla Dam. The bedrock, i.e. the Siwalik clays and siltstones, dip 10—15° to the northeast at Mangla, 45° at Jari, and are vertical south of the Changer Anticline (Fig. 78). During construction, it was found that shear zones extended for distances of the order of 500 m and varied in thickness from a few millimeters to over 1 m.

LAKE BENMORE, NEW ZEALAND

The physiography of Lake Benmore has been described by Adams (1974). It is the largest artificial reservoir of New Zealand and was impounded in December, 1964. The lake has a maximum depth of 96 m and a capacity of $2,040 \times 10^6$ m^3. The Benmore Dam is situated near 44.56°S, 170.20°E, about 85 km northwest of the town of Mamaru. Lake Benmore has been formed on the Waitaki River. This river drains the Lakes Ohau, Pukaki and Tekapo and the Mackenzie Basin to the plains of South Canterbury and North Otago. The lake has two arms, one extending west-northwest for about 15 km along the Ahuriri River and the other extending in a northerly direction for about 25 km (Fig. 79). Lake Aviemore was impounded in July 1968, some 15 km below Benmore Dam. This lake extends 10 km upstream and has a capacity about 23% that of Benmore. Another lake, Lake Waitaki which is situated further downstream, has been in existence since 1935. Its volume (40×10^6 m^3) is negligible in comparison to the newer lakes constructed upstream. The basement rock surrounding the lake consists of indurated graywacke and argillite of Triassic and Permian age.

According to Adams (1974), before the impounding of Lake Benmore only one major earthquake seems to have occurred in this area — on May 8, 1943. The earthquake has been assigned a magnitude of 6¼, an epicenter at 44.5°S, 169.9°E (about 25 km west-northwest of the Benmore Dam), and a focal depth of 70 km. Adams mentions that since the epicenter locations in the South Island at that time were very inaccurate, it is probable that the earthquake originated further to the west, near Lake Wanaka, where the highest intensities were reported. No new stations have been commissioned to investigate the seismicity of Lake Benmore. The closest stations are those at Mt. John (70 km away, installed in 1965), at Oamaru (80 km away, installed in 1965) and at Roxburgh (120 km away, installed in 1957 — see Fig. 80).

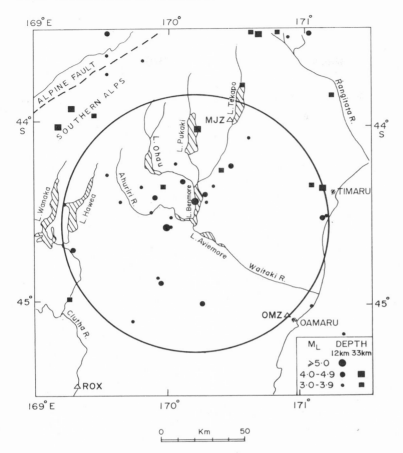

Fig. 79. Earthquakes located in the Benmore region for the pre-impounding period 1955—1964. The seismograph stations are shown by triangles; circle of 80 km radius is centered on the site of the Benmore Dam (after Adams, 1974).

Figs. 79 and 80 show the epicenters located in the Benmore region during the pre-impounding (1955—1964) and the post-impounding (1965—1972) periods. Adams (1974) mentions that there was no apparent seismic effect at the time of the impounding of Lake Benmore in December, 1964; however, statistical studies of routinely located earthquakes show a significant increase in seismic activity since then. The earthquake-occurrence frequency within 80 km of the dam has increased by a factor of 3—6, with a significance of 97—98%. A concentration of activity near the dam is suggested from the distribution of the post-impounding earthquakes. The activity falls off with distance, and beyond 60 km it remains constant (Fig. 81). After the impounding, earthquakes have mostly occurred in an area 40 km upstream of the dam. With the exception of a linear trend of earthquakes running in a

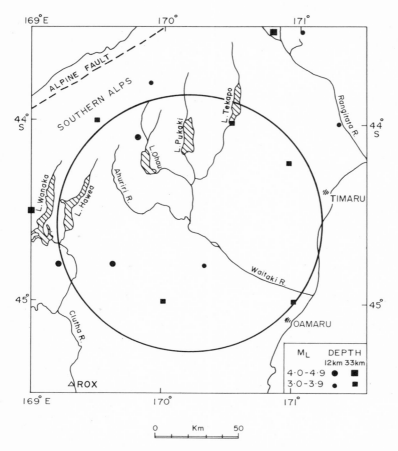

Fig. 80. Lake Benmore. Earthquakes located for the post-impounding period 1965—1972. The seismograph stations are shown by triangles; circle of 80 km radius is centered on the Benmore Dam (after Adams, 1974).

northeasterly direction from the north of the reservoir, there does not seem to be any clear correlation between the epicenters and the surface geology. The two largest post-impounding earthquakes had a magnitude of 5.0 and occurred within 20 km of the dam at time intervals of 1½ and 6¼ years after impounding.

KAMAFUSA DAM, JAPAN

A. Hasigawa (personal communication, 1975) reported a clear relationship between increased microearthquake activity and filling of the dam reservoir in Kamafusa, about 20 km west of Sendai, Japan. Microearthquake observa-

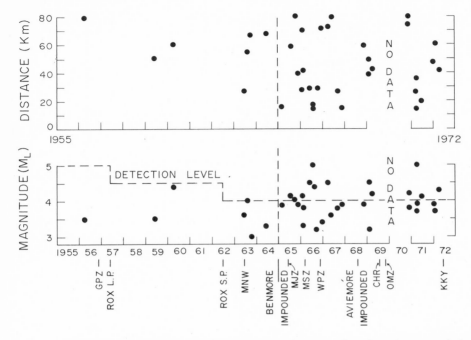

Fig. 81. Lake Benmore. Detection level and time history of earthquakes recorded within 80 km of the Benmore Dam. Magnitude and distance of earthquakes from the dam are indicated. Dates of reservoir impounding and installation of the seismograph stations are also indicated (after Adams, 1974).

tions were initiated in August 1963, six months prior to the filling of the reservoir, which commenced in February, 1970. The microearthquake activity increased in April, 1970, when the water level registered a maximum of 50 m. For the local microtremors (S − P ≤ 1.0 sec), the increase was about 100 times compared to the post-impounding microearthquake activity. Hasigawa mentioned that the change of activity which occurs with the seasonal variations in water level is very interesting. Every year, peak activity was observed during April and September, coinciding with the highest water levels. The hypocenters have been determined using a net of tripartite stations revealing a microearthquake zone 1.5 km northeast of the dam. The focal depths have been estimated to be within 3 km and the maximum magnitude registered has been 2.5.

A. Hasigawa (personal communication, 1975) also informed us of the following observations made by the staff of the Aobayama Seismological Observatory, Tohoku University, Sendai, Japan: (1) there is a NE-trending fault which passes through the dam site area; (2) there is a hot spring nearby in which the volume of discharge and the temperature increased after filling of the reservoir; (3) the rocks in the epicentral area consist of volcanic tuff;

and (4) the fracture coefficient of the rocks near the dam is higher than those for the nearby regions.

HSINFENGKIANG DAM, CHINA

Frequent.earthquakes occurred soon after the impoundment of the Hsin-fengkiang reservoir, situated 160 km northeast of Canton, China (Chung-Kang et al., 1974). The seismograph stations were timely set up and extensive research work was carried out. The dam was originally designed for intensity VI (MM scale). The dam was timely strengthened in 1961 to resist an earthquake of intensity VIII. Subsequently, on March 19, 1962, a strong earthquake of M_s = 6.1 and maximum intensity VIII occurred in the reservoir area. The dam withstood this strong earthquake, nevertheless some cracks on the dam appeared and warranted a second-stage strengthening. The research work of Chung-Kang et al. (1974) consisted of the study of the geological environment of the reservoir, features of seismic activity, and the study of aseismic behavior of the dam dealing with the characteristics of ground motions at the dam site and earthquake response of the dam. The salient features of their study are as follows.

The reservoir is lying on the huge E—W granitic mass of Late Mesozoic age. To the east of the reservoir lies a long and narrow fault basin, with Tertiary formations trending NE. To the south and north of the reservoir, Upper Paleozoic to Jurassic and Cretaceous strata form extensive mountainous areas. Here various volcanic eruptions occurred during the Upper Jurassic and basalt flows during the Early Quaternary. The reservoir area has been unstable since Mesozoic and Cenozoic times, when various faults and fold systems were formed. The focal mechanism of the main earthquake mainly indicated a strike-slip motion in conformity with the ENE-trending faults.

There is no historical record of destructive earthquakes in this region before the impoundment of the reservoir, although, four felt earthquakes of intensity V—VI occurred in the 25-year period prior to impoundment. The Canton seismograph station, situated 160 km southwest of the dam started recording earthquakes in the reservoir area from October, 1959, onwards, one month after the beginning of the impoundment. As the water level rose rapidly, the earthquakes became more frequent. The network of seismograph stations established in July, 1961, recorded 258,267 shocks of magnitude M_s ≥ 0.2 up to December, 1972, out of which 23,513 shocks had a magnitude M_s ≥ 1.0. The important observations on distribution of seismic activity in the reservoir area are: (1) most of the shocks occurred near the intersections of major faults and in rock masses with interbedded weak layers; (2) the seismically most active areas and strongest shocks were located in the vicinity of the dam where the water depth was deepest (80 m); and (3) the focal depths were very shallow, i.e. from 1 to 11 km.

The seismic characteristics found by Chung-Kang et al. (1974) are as given below.

(1) The earthquake sequence consisted of a mainshock with numerous foreshocks, and aftershocks decreasing slowly.

(2) The *b* value in the frequency—magnitude relationship is similar to those for other reservoir-induced earthquake sequences but different from that of the normal tectonic earthquakes. The *b* value of the foreshocks (1.12) is higher than for the aftershocks (1.04). These values are 1.4—1.5 times higher than those for the tectonic earthquakes (0.72).

(3) The magnitude ratio of the largest aftershock to the mainshock is high (0.87).

(4) The number of aftershocks attenuates slowly with an attenuation coefficient 0.9.

Chung-Kang et al. (1974) state that a close relationship existed in time and space between reservoir filling and the earthquakes. The seismic activity was affected by changes in water level. A rapid rise of the water level to a high level was often followed by an increased seismicity. Chung-Kang et al. suspect that the groundwater channels of deep circulation may have formed in the fissured zones and in the accompanying transverse faults lying near the large faults. According to them, with the rise of seepage pressure, the groundwater gets a deeper circulation and causes (a) the decrease in normal stress on the fault, and (b) the argillization of the rocks of the weak structural plane and thus reduction of its shearing strength. In the Hsinfengkiang reservoir, as the water level rose to 20 m, percolation of water began to take place along the NNW fault zone on the northwest side of the dam to induce minor earthquakes. When the water level rose to 50—60 m, the seepage pressure increased to cause a series of small earthquakes and stronger earthquakes of magnitude $M_s \geqslant 3.0$. When the water level reached its peak level, the seepage pressure became highest accompanied with an upsurge of seismic activity and the occurrence of the strongest earthquake.

OTHER SITES

In addition to the previously discussed cases where the initiation or enhancement of the seismic activity following reservoir impounding or fluid injection has been evidenced, there are many other areas showing similar effects in a small way. We mention some of them here.

Lake Eucumbene, situated in the Snowy Mountains of Australia, has a capacity of $4{,}170 \times 10^6$ m^3 and the dam height is 116 m. The lake was impounded in 1970 and Timmel and Simpson (1973) mention that a number of tremors, including one of magnitude 5, were recorded during and after filling by a network of seismograph stations which began operating after the filling of the lake commenced. Since little is known about the seismicity of the area in the time prior to the filling, it cannot be said with certainty whether the impoundment of the reservoir enhanced the seismic activity. An earthquake has also been reported from Lake Blowering, also situated in the Snowy Mountains area of Australia, following impounding. The lake has a capacity of $1,718 \times 10^6$ m^3, the dam height is 113 m, and it was impounded in 1970.

Rothé (1968, 1970) pointed out a close correlation between reservoirs north of Lerida, Spain, and the shocks in the vicinity of the reservoirs. Of particular interest is the earthquake of June 9, 1962, which had an intensity V (MM scale) and whose epicenter (42.0°N, 0.6°E) coincided with the position of the Canalles Dam (Fontsere, 1963). The Canalles Dam is an arch-type dam and has a height of 150 m.

The Camarillas reservoir in Spain triggered seismic activity in 1961 when it was filled to about 30 m height. The dam is built on a limestone diapir. The diapir structure is probably due to an intrusion at depth. According to

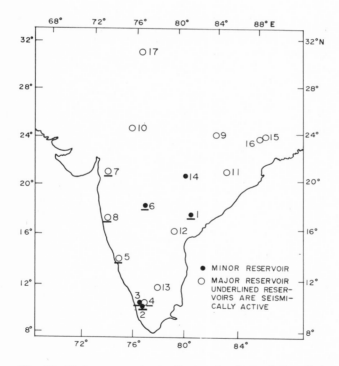

Fig. 82. Water reservoirs in Peninsular India (after Guha et al., 1974). The underlined reservoirs are seismically active. 1 = Kinnersani, 2 = Sholayar, 3 = Mangalam, 4 = Parambi-kulam, 5 = Sharavathi, 6 = Ghirni, 7 = Ukai, 8 = Koyna, 9 = Rihand, 10 = Rana Pratap Sagar, 11 = Hirakud, 12 = Nagar Juna Sagar, 13 = Mettur, 14 = Itiadom, 15 = Maithon, 16 = Panchet, and 17 = Bhakra.

Lomnitz (1974), the added hydrostatic head in the reservoir might have reduced the strength of the basement rocks sufficiently for failure.

Guha et al. (1974) have considered the seismic status of seventeen reservoirs in India. Eight of the reservoirs have evidenced some seismic activity. Reservoirs like the Kinnerasani, Parambikulum, Sharavathi, Bhandara, and Ukai Dams had definite seismic activity following impoundment (Fig. 82). In the cases of other reservoirs, like Ghirni, Mangalam, Sholayar, etc., Guha et al. (1974) reported the occurrence of isolated shocks. Out of all these cases, a special study was undertaken near the newly impounded Mula reservoir (19°23'N, 75°39'E) which is situated on Deccan Traps. An ultrasensitive seismograph was used to record ultra microearthquakes with a magnitude below zero. The seismograph has a magnification of 10^6 operating in the frequency range of 10—70 Hz. Numerous microearthquakes with a magnitude below zero having ground motions of 20—50 Hz were recorded. These earthquakes occurred at a shallow depth of less than 1 km. Fig. 83 shows that the frequency of these microearthquakes increased significantly with the rise in water level of the reservoir. Prior to the formation of the reservoir the area was free from seismic activity.

Roksandic (1970) reported an increase in the seismic activity following the filling of the Bileca reservoir in Yugoslavia.

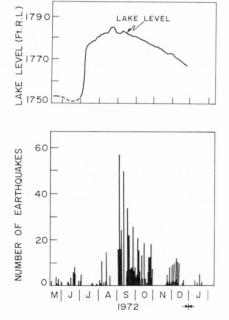

Fig. 83. Mula reservoir. Lake levels and frequency of microearthquakes (magnitudes between —2 and 0) (after Guha et al., 1974).

According to Mickey's report (1973a, b), in the United States there are 52 dams built, 7 under construction and 9 in the planning stage with a dam height of over 100 m. In addition to studying the seismicity in these areas, he also investigated earthquake occurrences near several other dams of heights less than 100 m which are situated in seismic zone 1 and have historical reports of seismic activity in the immediate vicinity. Ten out of the eighteen reservoirs where recorded seismic data are available show a possible cause and effect relationship. Indirect and inconclusive evidence is available relating seismic activity and reservoir impounding for more than 40 instances. This evidence is found in felt-earthquake reports which appeared in newspapers, records compiled from interviews, and information derived from earthquake enquiry cards. Mickey quoted some examples of suspected reservoir-related earthquakes from the annual United States earthquakes reports of the U.S. Coast and Geodetic Survey. According to these reports, nearly 200 shocks reported in 1955 from eastern Washington were attributed by geologists to the settlement of subterranean rock under the weight of irrigation water. In 1961, several shocks felt in Washington were attributed to the impounding of water in the newly-created Lake Entail behind the Rocky Reach Dam. Similar reports were made in connection with the Shasta Dam (Shasta Lake), California, in 1944, the Kerr Dam (Flathead Lake) Montana, where over 100 quakes were felt in 1969, and several other dams.

The seismicity associated with the Mississippi River Valley, Missouri, also needs to be mentioned. McGinnis (1963) has reported about 70 earthquakes of intensity V which occurred in the Mississippi region since 1811. The epicenters are concentrated near the river. He found that the earthquake activity increased when the rate of change in the river stage increased, and that the rate of the energy release rose with the water load. McGinnis regarded the water load as aiding the movement of a subsiding crust.

Caloi (1970) has indicated that alternating increases and decreases of the water level of the Pieve di Cadore Lake in Italy were followed by a clear increase in seismicity. After having reached a maximum level in October, 1963, the lake was drained, and by the end of March, 1964, it was virtually empty. This was followed by a series of microshocks. Between March 5 and 26, 1964, some 17 microshocks were recorded. From May to August in that year a total of 65 microshocks were recorded. The five strongest shocks which had a magnitude of 1.5—2 and which were felt locally, were within, or very close to, the reservoir area. The other shocks took place over a relatively wide area around the lake. The Pieve di Cadore Dam is founded on dolomite rocks of Upper Triassic age which have developed almost perpendicular to the river. Fractures caused by tectonic movements have been observed in the reservoir area.

STATISTICAL ANALYSIS OF RESERVOIR LEVEL AND TREMOR FREQUENCY
DATA

Statistical analyses of water-level and earthquake-frequency data have
been carried out for a few cases. In Fig. 84, the average monthly water level
and number of earthquakes recorded per three-month period are shown from

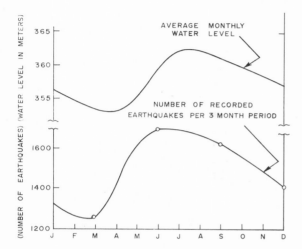

Fig. 84. Lake Mead. Correlation plot of average monthly water level and number of
earthquakes recorded per three-month period at Lake Mead, from 1939 through 1951
(after Mickey, 1973a).

1939 to 1951 for Lake Mead. In Fig. 85, the three monthly average of water
levels and the earthquake frequency as recorded per three-month period are
given and plotted against the middle month for the whole of the periods
considered for Koyna, Kariba and Kremasta (Gupta and Rastogi, 1974b).
The correlation between the water level and the earthquake frequency is
very apparent from these figures. A correlation coefficient of +0.94 could be
obtained for Lake Mead's reservoir level and earthquake frequency. For
Koyna, Kariba, and Kremasta the correlation coefficients were found to be
+0.93, +0.74, and +0.69, respectively. Although the values of the correlation
coefficients are quite large, they have a low level of significance. Moreover,
the high activity before and after the mainshock greatly affects the frequency curve and when using other averages for water level and different
time intervals for the earthquake frequency the correlation is much less
apparent.

Hagiwara and Ohtake (1972) have performed the cross-correlation analysis
between the number of shocks in the ith month, n_i, and the water level, with

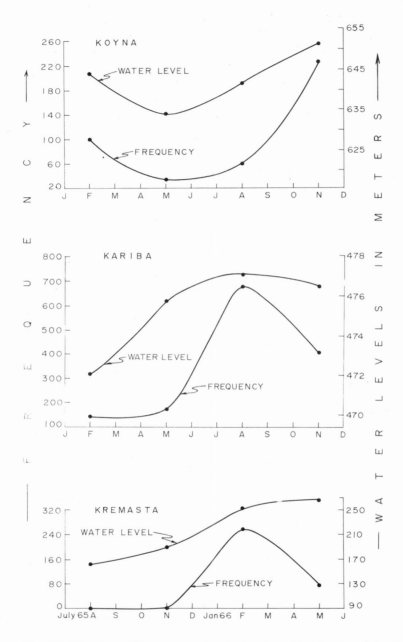

Fig. 85. Three monthly average of water levels and totals of earthquake frequency for the same months at Koyna, Kariba and Kremasta; the correlation coefficients are found to be +0.93, +0.74, and +0.69, respectively. The periods include 1964—1968 for Koyna; 1959—1968 for Kariba; and July, 1965, to June, 1966, for Kremasta.

a timelag of τ months $l_{i+\tau}$, for the Kurobe reservoir, using the formula:

$$r_\tau = \frac{\sum_i (n_i - \bar{n})(l_{i+\tau} - \bar{l}_\tau)}{\left[\sum_i (n_i - \bar{n})^2 \sum_i (l_{i+\tau} - \bar{l}_\tau)^2\right]^{1/2}}$$

where $\bar{n}$ and $\bar{l}_\tau$ are the mean of n_i and $l_{i+\tau}$, respectively. The highest correlation coefficient, +0.41, has been found for a zero time lag indicating that the frequency of the dam-site shocks is positively correlated to the water level, with a time lag, if any, of less than 1 month. The value of correlation coefficient is significant.

Healy et al. (1968) have calculated the cross correlation between earthquake frequency and fluid pressure to determine the time lag between the peak fluid pressure and peak seismicity. The formula used was:

$$R_{XY}(p) = \frac{1}{n-p} \sum_{q=1}^{n-p} X_q Y_{q+p}, \quad p = 0, 1, 2, \ldots m$$

where X is the data from earthquake frequency, Y is the data from the pressure series, $R_{XY}(p)$ is the cross covariance or cross correlation, n is the number of discrete data points, and p is the time lag. For treating the data gaps in the pressure series the method of Lee and Cox (1966) has been used. From the above formula it was found that seismic activity followed the peak pressures with a time lag of 10 days. A maximum correlation coefficient of +0.5 was obtained for this lag.

Factors affecting the tremor frequency

In the cases discussed, it is seen that seismicity generally increases with the rise in reservoir water level/rate of injection in a well. In a few examples the correlation looks fairly good, although statistically it is not significant. From the examination of these cases it has been inferred by Gupta et al. (1972a) that, in addition to the tectonic setting and geological conditions, among the factors affecting the tremor frequency near the reservoirs are: (1) the rate of increase of water level; (2) the duration of loading; (3) the maximum load reached; and (4) the period for which high levels are retained.

The importance of the rate of loading has been demonstrated theoretically by Snow (1968b) and others. A rise in the water level is in general followed, after a certain time lag, by an increase in seismicity. The strongest earthquakes occurred after a rapid rate of loading. The increase in seismic activity and the time lag depend on the above-mentioned four factors. Howells (1973) assessed the time required for a substantial increase of pore pressure at varying depths consequent to a surficial increase of pore pressure. Using the one-dimensional diffusion equation, he has estimated that the time taken by surface pore pressures to be transmitted to a depth of 5—10 km is of the order of 100 days.

Snow (1972) has observed that the onset of earthquakes occurs at a higher threshold as time passes. The threshold level increases with time due to seismic stress releases or plastic flow. In the Kremasta area, seismic activity commenced when the level reached an elevation of 245 m, from tail-water elevation of 140 m, during 1966. The maximum seismic activity occurred when the elevation was about 268 m. In later years, the tremor frequency did not increase below the 270-m level. In June 1971, when the level was only about 3 m below maximum, no shocks were felt. However, shocks were felt (several per week) when the level rose during the rainy season to 270 m and higher.

In the case of the Koyna reservoir, between 1963 and 1967, whenever the water level exceeded the 2,140-ft (652 m) mark and was retained at a high level for a long time, seismic activity increased considerably after a certain time lag. During 1967, when the water level reached the 2,154-ft (657 m) mark and was retained for 132 days, maximum seismic activity resulted, including the two largest earthquakes experienced in the region. During 1968, the maximum level was 2,120 ft (646 m), during 1969 it touched the 2,140-ft (652 m) mark, and during 1970 and 1971 it was about 2,145 ft (654 m). The seismic activity in these years has not been significant, except for the activity of October 1968 when a magnitude 5 earthquake occurred. In 1973, when the lake was filled to its capacity and the water level touched the 2,159-ft (658 m) mark, the seismic activity increased significantly and included an earthquake of a magnitude exceeding 5. This level is the maximum which the Koyna reservoir can attain, and hence it is quite likely that in subsequent years, even if the reservoir attains this level, the activity may not increase significantly if the threshold level, as discussed above, has been raised. From the cases of reservoir-associated seismicities examined in this chapter, it appears that the seismic activity decreases with time and terminates after a couple of years.

The reservoir sites which showed significant seismic activity after impounding were either aseismic or had only occassional tremors before impoundment. Post-impounding seismicity has been confined to the reservoir's immediate vicinity in most cases. In some cases, seismic activity spreads away a little from the reservoirs, but the epicenters were mostly located within 25 km or so from the reservoirs. At the seismically active sites, an increase in seismicity from background level has been recorded after impounding; however, this cannot be established without doubt, e.g. in the Mangla and Benmore areas.

As the geological factors near the reservoirs showing increased seismicity are of diverse character, it is not possible to generalize them. Nevertheless, certain features are common among some of the examples. The necessity of hydraulic continuity to deeper layers in inducing the earthquakes in the reservoir areas has been demonstrated by the Lake Mead and Dale examples. In the Virgin-Detrital Basin of Lake Mead, the hydraulic continuity between the reservoir water and deeper layers is probably barred by impervious deposits of salt and clay, a few thousand meters thick. A number of faults, which are present beneath this basin, were probably not activated in the absence of hydraulic continuity. The seismic activity of the Lake Mead area is mostly confined to the Boulder Basin where salt deposits have not been found. Similar observations have been made for the high-pressure injection wells in Dale, New York. During 1971, in a high-pressure injection well for hydraulic salt mining, fluid injection was performed at the basal contact of the salt bed and dolomite. This well was connected to a series of recovery wells which bottomed in, or very close to, the Clarendon—Linden Fault zone, and it seems that the injected fluid reached this active fault zone and triggered the release of tectonic strain. The tremor frequency very close to the fault increased from one event every few months to 100 events per day. When the well was shut down, the frequency decreased to the pre-injection level within 2 days. In 1972, when injection under pressure in a well, 0.3 km north of the previous well, was confined to the salt bed, no increase in tremor frequency was noticed. In 1972 the water loss was negligible while it was high in 1971. Hydraulic continuity to deeper layers has also been inferred in the cases of Koyna, Kariba and Kremasta. The presence of hot springs in these areas is an indication of the deep circulation of water. Faults and fractures are present near most of the reservoirs which permit the flow of water below and away from the reservoirs. In some of the regions, permeable rocks like sandstones are present, and in other cases less permeable rocks are present. In the Koyna area, the vesicular traps are permeable. Limestones which are easily affected by water, and which, at some places, developed karst features, are widespread in several regions such as Kremasta, Vajont, Monteynard, Oued Fodda, and others. The limestones play an important role in flow of water. Many of the reservoir sites have a volcanic history. Volcanic rocks are present in areas such as Koyna, Kariba, Hendrik Verwoerd, Marathon, and Lake Mead.

Chapter 4

CHARACTERISTICS OF RESERVOIR-ASSOCIATED EARTHQUAKES

Investigations of the characteristic features of reservoir-associated earthquake sequences are useful in understanding the part played by reservoirs in causing these earthquakes. The characteristic features studied for these earthquakes are: the slope in the frequency—magnitude relation, the magnitude ratio of the largest aftershock to the mainshock, foreshock—aftershock patterns and the time distribution of foreshocks and aftershocks. These characteristic features of earthquakes reflect upon the mechanical structure of the media and the nature of the applied stress. The discussion in this chapter shows that the reservoir-associated earthquakes have different characteristics as compared to normal earthquakes (not associated with reservoirs) of the region concerned, from which it has been inferred that the artificial lakes are responsible for changing the mechanical properties of the strata.

Focal mechanisms of the reservoir-induced earthquakes and earthquakes triggered by fluid injection have been also discussed in order to comprehend the part played by reservoir filling and fluid injection in causing such earthquakes.

THE FREQUENCY—MAGNITUDE RELATIONSHIP

The first investigation of the frequency of earthquakes in relation to their size was made by Ishimoto and Iida (1939). They studied the frequency of occurrence of maximum trace amplitudes on seismograms for earthquakes in the Kwanto region of Japan. Gutenberg and Richter (1954) studied the magnitude and frequency distribution of large earthquakes for the whole world as well as for certain specific regions. Subsequently, detailed regional investigations for various areas have been made by several authors and it has been found that, in general, the frequency distribution of earthquakes over an observed range of magnitudes in a particular area can be represented by a simple relation (Richter, 1958) of the type:

$$\log N = A - bM$$

where N is the number of shocks of magnitude $\geqslant M$, A and b are constants, and logarithms are taken to the base 10. The value of A depends upon the period of observation, the size of the region considered and the level of seismic activity, whereas b depends upon the ratio of the number of earthquakes in the low- to high-magnitude groups.

Detailed statistical calculations on the frequency—magnitude relations of the earthquakes have been also carried out by Utsu (1965). According to him, b values could be obtained empirically by the following relation:

$$b = \frac{0.4343m}{\sum_{i=1}^{m} M_i - mM_{min}}$$

where m is the total number of earthquakes and M_{min} is the lowest magnitude considered.

The parameter A is found to vary significantly from region to region, whereas the parameter b does not vary much. Values of b have been reported from about 0.5 to 1.5, mostly lying between 0.7 and 1.0 (Isacks and Oliver, 1964). The parameter b has been investigated by several seismologists. Some believe that this parameter is constant and equal to about 1.0 and that the different values obtained by investigators are due to the difference in data and methods used for computations. Nevertheless, a majority believes that b varies from region to region and with focal depth and that its value depends on the stress conditions and on the heterogeneity of the rock volume generating the earthquakes.

Riznichenko (1959), Bune (1961) and some others found relatively uniform worldwide b values. On the other hand, Gutenberg and Richter (1954), Miyamura (1962), Tomita and Utsu (1968), Karnik (1969), Evernden (1970) and others have shown regional variations in the values of b and have emphasized their possible tectonic significance. A decrease of b with an increase in focal depth has been found by Karnik (1969), Papazachos et al. (1967), Brazee (1969), and others.

Mogi (1962b, 1967a) has examined the values of b in a laboratory by studying the brittle fracture of rock specimens. He found that b depends upon the mechanical heterogeneity of the rock samples and increases with the increase in heterogeneity. On this basis, he inferred that the values of b of shallow earthquakes are related to the mechanical structure of the earth's crust. Later, Mogi (1967b) indicated that b may not be very sensitive to the structure of the earth's crust since it falls in a narrow range of 0.6—1.0 for most of the regions and the mechanical structure of the earth's crust may vary significantly from region to region, except in some volcanic or highly fractured regions. Scholz (1968) found that b depends on the percentage of the existing stress within the rock sample to the final breaking stress.

There are many uncertainties in determining the values of b and in their comparison from region to region (Allen et al., 1965). Measurements of both magnitude and number of earthquakes have more or less random errors. Uncertainties in magnitude measurements are sometimes several tenths of a unit. These random errors in number and magnitude cause the scattering of points in a plot of the observed data. The systematic errors are not readily apparent. Such errors are largely caused by the different magnitude scales used. The systematic errors could become significant for the regions where extensive data is not available for constructing a magnitude scale. For finding the *b* value graphically, two methods of representing the earthquake frequency are in vogue. The first method is to plot the number of earthquakes with a given increment of magnitude (generally 0.1), while the second method is to plot the number of earthquakes equal to or greater than a given magnitude. The second method using a cumulative number gives a smoother plot which is not affected by the magnitude interval.

A linear relation between log *N* and *M* has been found to be valid over a wide range of magnitudes. But in many cases, since the number of earthquakes which are missed by a seismic network increases with decreasing magnitude, the cumulative frequency distribution curve becomes horizontal

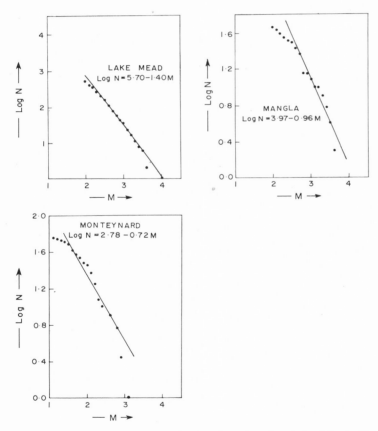

Fig. 86. (a) Frequency—magnitude plots for the Lake Mead, Mangla, and Monteynard sequences.

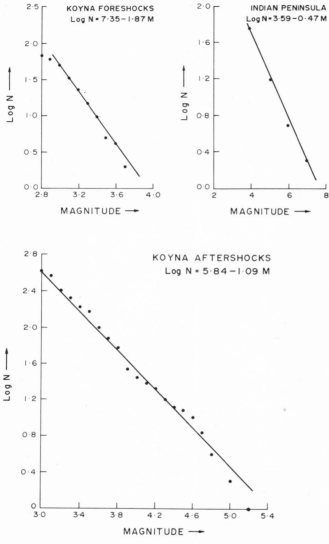

Fig. 86. (b) Frequency—magnitude plots for Koyna foreshocks and aftershocks and for the Indian Peninsula.

in the lower-magnitude range. This poses the problem of deciding the lower magnitude limit in the b-value determination for a particular set of earthquake data. A change in this lower limit would give a different estimate for b for the same set of data. By limiting the sample of earthquakes analyzed to events bigger than a certain magnitude, larger and more reliable estimates of b are obtained, but at the same time the number of earthquakes in the

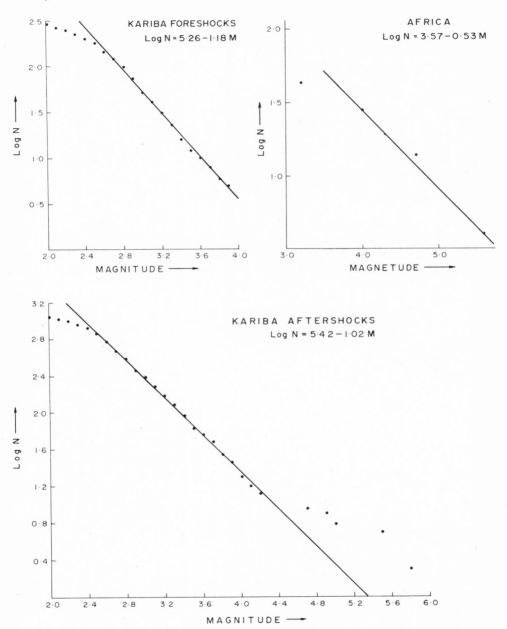

Fig. 86. (c) Frequency—magnitude plots for the Kariba foreshocks and aftershocks and for the African region. For the Kariba aftershocks, data in the magnitude range 4.7—5.8 does not fit in the linear relationship.

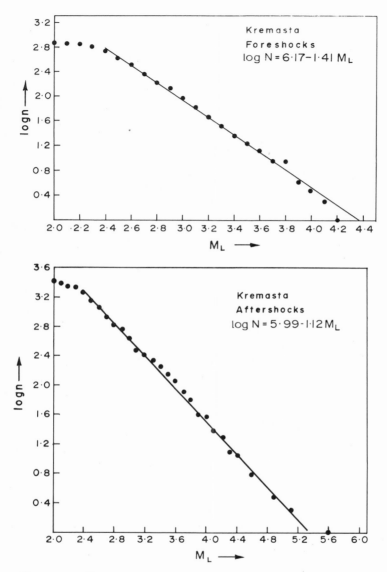

Fig. 86. (d) Frequency—magnitude plot for Kremasta foreshocks and aftershocks (after Comninakis et al., 1968).

sample rapidly decreases and the uncertainty in b correspondingly increases. In making a regional comparison of b, it is necessary to take into account the fact that different regions have different detection capabilities for earthquakes.

TABLE VIII

The *b* values for reservoir-associated earthquake sequences and normal earthquakes in the regions concerned

Region and time interval	Number of earthquakes	Magnitude range	*b*
Lake Mead 1941—1942	536	2.0—4.0	1.40
Monteynard Dam 25-4-1963 to 13-11-1967	57	1.1—3.1	0.72
Mangla reservoir 28-1-1966 to 17-11-1968	46	2.0—3.6	0.96
Kurobe Dam	110	2—25 mm amplitude	1.46
Kariba reservoir foreshocks 8-6-1959 to 23-9-1963	291	2.0—4.0	1.18
Kariba reservoir aftershocks 23-9-1963 to 27-12-1968	1,114	2.0—5.8	1.02
Africa region 1-1-1963 to 30-6-1966	43	3.2—5.6	0.53
Kremasta reservoir foreshocks 1-9-1965 to 5-2-1966	740	2.0—4.2	1.41
Kremasta reservoir aftershocks 5-2-1966 to 30-11-1966	2,580	2.0—5.6	1.12
Normal earthquakes in the Kremasta region			0.64
Greece region — shallow normal earthquakes			0.82
Koyna reservoir foreshocks 10-9-1964 to 13-9-1967	51	2.8—3.7	1.87
Koyna reservoir aftershocks 10-12-1967 to 27-6-1969	422	3.0—5.2	1.09
Godavari Valley sequence 13-4-1969 to 2-5-1969	52	2.1—5.7	0.51
Indian Peninsula earthquakes from historical records of more than 300 years	52	4.0—7.0	0.47

The log *N* versus *M* relationships for the reservoir-associated earthquake sequences are shown in Fig. 86. The *b* values for these sequences, along with the regional *b* values obtained from normal earthquakes, are listed in Table VIII. The second decimal place in the *b* values is not important, however, it

TABLE IX

Source of the data used for Table VIII and magnitude scales

Richter scale (M_L)	Local magnitude scale equivalent to the Richter scale	Other magnitudes converted to the Richter scale
Mangla. Data from Adams (1969)	*Lake Mead.* Data from Jones (1944)	*Africa.* The m_b values given by Mizoue (1967) converted to M_s using Richter's (1958) relation $M_s = 1.59\ m_b - 3.97$. M_s is equivalent to M_L
Koyna. Data from Guha et al. (1968, 1970).	*Godavari Valley sequence.* The b value obtained by Gupta et al. (1970). The magnitudes calculated from surface waves are equivalent to Richter scale magnitudes	*Indian Peninsula.* The intensity values (I) given by Gubin (1969) converted to M_L using Gutenberg and Richter's (1956) relation $M_L = 1 + \frac{2}{3}\ I$
	Kariba. Data from Archer and Allen (1969). The magnitude values in this catalogue are equivalent to M_L as mentioned by Gough and Gough (1970b)	
	Kremasta. The b values obtained by Comninakis et al. (1968)	
	Kremasta area normal earthquakes. The b value obtained by Galanopoulos (as quoted by Comninakis et al., 1968)	
	Greece region. The b value obtained by Galanopoulos (1965)	

has been retained in view of the other works cited. An attempt has been made to remove, as much as possible, the earlier-mentioned uncertainties in these determinations. Most of the magnitude values have been changed to the Richter scale or an equivalent local scale, if they were otherwise (Table IX). The earthquakes used in these determinations fall in the small- to moderate-magnitude range. Most of the values have been obtained graphically, taking the earthquake frequency as the cumulative number. The remaining uncertainties are expected to be within reasonable limits, and hence it seems to be reasonable to compare the b values of different sequences as given in Table VIII.

When comparing the figures in Table VIII, the following needs to be mentioned. For the Kurobe earthquakes in Japan, Hagiwara and Ohtake (1972) have determined the slope, m, of the frequency versus log-amplitude curve to be 2.46. The m value is converted to the b value by using the relation $b = m - 1$ (Mogi, 1963a). For the Monteynard sequence, four magnitude values greater than 4.0 do not fit in the linear relationship. For Mangla, the cumulative frequency curve tends to become horizontal for a magnitude less than 2.0 and, hence the data below this magnitude has not been considered in obtaining the b value. For the Koyna, Kariba and Kremasta sequences, earthquakes occurring before the mainshock have been considered as foreshocks and earthquakes occurring after the mainshock as aftershocks.

For the reservoir-associated earthquake sequences the b values are mostly $\geqslant 1$. These values are relatively higher than normally found for such small magnitude ranges considered in these cases. Moreover, the foreshock b values of the reservoir-associated earthquakes are higher than the aftershock b values; also both these values are higher than those for the normal earthquake sequences of the regions concerned. This fact discriminates the reservoir-associated earthquakes from the normal earthquakes.

Berg (1968) has reviewed the foreshock and aftershock b values for earthquakes in Japan, Alaska, Greece, and Chile. He found that the foreshock b values are much lower (0.3—0.6) than the aftershock b values (0.75—1.2). He also found that a high b value ($\geqslant 0.5$) for a foreshock sequence is an indication of the occurrence of a large-magnitude earthquake. Considering the experimental results of Mogi (1962a, b) and Scholz (1968), Berg has argued that low b values are associated with high stress and strength, whereas high b values are associated with reduced strength and low stress after a major earthquake. However, in the three significant cases of reservoir-associated earthquake sequences, results differing from Berg's analysis have been obtained. The foreshock b values determined for the Kariba, Kremasta, and Koyna sequences are 1.18, 1.41, and 1.87, while the corresponding aftershock b values are 1.02, 1.12, and 1.09, respectively. This probably is also evidence of the influence of the impounding of the lakes on the manner of foreshock occurrence. However, this evidence must be considered weak because not many well-recorded ordinary foreshock sequences are available.

To test the statistical significance of the above-mentioned differences in the b values of the earthquake groups, Gupta and Rastogi (1974b) carried out an analysis similar to Utsu (1966), using the distribution, F. Let S_F and S_A denote the total number of foreshocks and aftershocks, and b_F/b_A is calculated and compared with the F value for $2S_A$ and $2S_F$ degrees of freedom at a particular confidence level, which can be found in most books on statistics. In this analysis, the F-value tables given by Bowker and Lieberman (1959) have been used. The results of the analysis are given in Tables X and XI. Table X shows that the differences in foreshock and aftershock b values for the Koyna, Kariba and Kremasta sequences are significant at the 95% confidence level since the b_F/b_A values are greater than the F values read at the 95% confidence level. Similarly, it can be seen from Table XI that the difference between the b values for the Koyna aftershocks and those of the

TABLE X

Statistical significance test for the difference in foreshock and aftershock b values

		Koyna	Kariba	Kremasta
Total number of foreshocks S_F		51	291	740
Total number of aftershocks S_A		422	1,114	2,580
b for foreshocks	b_F	1.87	1.18	1.41
b for aftershocks	b_A	1.09	1.02	1.12
Degrees of freedom	$\nu_1 = 2S_A$	844	2,228	5,160
Degrees of freedom	$\nu_2 = 2S_F$	102	582	1,480
	b_F/b_A	1.71	1.16	1.26
F value at 95% confidence level		1.30	1.12	1.07

Since the F values are smaller than the b-value ratios, the difference in fore- and after-shock b values are significant at the 95% confidence level

Godavari Valley aftershocks and the regional b value of Peninsular India is significant at the 95% confidence level. Similar analyses for the Kariba and Kremasta regions could not be undertaken due to the lack of data.

TABLE XI

Statistical significance test for the differences in b value of Koyna aftershocks with those for the Godavari Valley sequence and Indian Peninsula earthquakes

Total number of Koyna aftershocks	S_K	= 422
Total number of Godavari Valley aftershocks	S_G	= 52
Total number of earthquakes in Peninsular India	S_I	= 52
b for Koyna aftershocks	b_K	= 1.09
b for Godavari Valley aftershocks	b_G	= 0.51
b for Peninsular India earthquakes	b_I	= 0.47
	b_K/b_G	= 2.14
	b_K/b_I	= 2.32
Degrees of freedom	ν_1	= 104
	ν_2	= 844
	$F_{0.05(104,724)}$	= 1.27

Since the F value is smaller than the two ratios of the b values, differences between the b value of Koyna aftershocks with those for Godavari Valley aftershocks and Indian Peninsula earthquakes are significant at the 95% confidence level

RELATIONSHIP BETWEEN THE MAGNITUDES OF THE MAINSHOCK AND THE LARGEST AFTERSHOCK

It has been observed that the magnitude, M_1, of the largest aftershock is related to the magnitude, M_0, of the mainshock. According to Bàth's (1965) law for large, shallow earthquakes:

$$M_0 - M_1 = 1.2$$

Papazachos (1971) found this law to remain valid from a study of 216 aftershock sequences with $M_0 \geqslant 5$ which occurred in Greece.

In Table XII, M_0 and M_1 values for reservoir-associated earthquake sequences are given. The following needs to be clarified for this table. Adams (1969) has listed the earthquake data for the Mangla sequence, which resembles a swarm-type of activity where no mainshock is clearly distinguishable. For such a case $(M_0 - M_1)$ will be very small. In the Mangla sequence, two shocks have the largest assigned magnitude of 3.6. However, they were not followed within a few days time by any aftershocks, whereas another shock of magnitude 3.5, which occurred on April 25, 1966, was followed on April 29 by aftershocks with a maximum magnitude of 3.3 and hence has been used in Table XII. Another earthquake of magnitude 3.5, which occurred on December 10, 1967, in Mangla, was followed by a shock of magnitude 3.0 on December 11.

Using the data given in Table XII for six of the reservoir-associated earthquake sequences, Papazachos (1974) found the following relationship

$$M_0 - M_1 = 0.6$$

with a standard deviation of 0.3. This relationship differs significantly from the one given by Bàth for the shallow, large, normal earthquakes. Reliable

TABLE XII

The magnitudes of the mainshock and largest aftershock and the b values for reservoir-associated earthquake sequences

Region	Mainshock magnitude (M_0)	Largest after-shock magnitude (M_1)	$M_0 - M_1$	M_1/M_0	b
Lake Mead	5.0	4.4	0.6	0.88	1.40
Monteynard	4.9	4.5	0.4	0.92	0.72
Mangla	3.5	3.3	0.2	0.94	0.96
Kariba	6.1	6.0	0.1	0.98	1.02
Kremasta	6.2	5.5	0.7	0.89	1.12
Koyna	6.0	5.2	0.8	0.83	1.09

magnitude data are not available for other known cases of reservoir-associated earthquake sequences, but they are known to exhibit a swarm-type of activity. Thus it is seen that $(M_0 - M_1)$ is much smaller for reservoir-associated earthquakes.

Utsu (1969) found a positive correlation between $(M_0 - M_1)$ and the b value of the sequence. For some sequences he found small b with small $(M_0 - M_1)$ values and large b with large $(M_0 - M_1)$ values. However, for reservoir-associated sequences we see large b with small $(M_0 - M_1)$ values. Instead of the difference $(M_0 - M_1)$, the ratio of the two magnitudes M_1/M_0 can also be considered. Utsu's observations would then mean that for a large b value, M_1/M_0 is small, while for a small b value this ratio is large. A similar observation was made by McEvilly and Casaday (1967) and McEvilly et al. (1967) for Californian earthquake sequences. They found that for the sequences with smaller b values (0.4—0.5) the M_1/M_0 is large (about 0.9) and for the sequences with higher b values (0.6—0.8), M_1/M_0 is small (0.6—0.7). Quite contrary to this, the reservoir-associated sequences have high b values along with a high M_1/M_0 ratio (Table XII). It is interesting to point out here that the Godavari Valley earthquake sequence of April, 1969, occurring in the same Deccan Shield of India as the earthquake at Koyna, had $M_0 = 5.7$, $M_1 = 5.0$ and thus $M_1/M_0 = 0.9$, while its b value is low, being 0.51. The same is true for the Broach earthquake sequence of March, 1970, which had a high M_1/M_0 ratio of 0.8 ($M_0 = 5.7$, $M_1 = 4.6$) and a low b value of 0.4 (Gupta et al., 1972c).

Since the difference between the magnitude of the mainshock and the magnitude of the largest aftershock depends on the stress conditions and the heterogeneity of the material in the seismic region (Mogi, 1963b), the above observations indicate that artificial lakes probably affect the stress distribution and the mechanical properties of the strata. However, evidence that the reservoir changes the mechanical property of the strata has been considered weak because large b with small $(M_0 - M_1)$ values or large M_1/M_0 values are also seen for some natural earthquake sequences (Papazachos et al., 1967; Chaudhury and Srivastava, 1973).

TIME DISTRIBUTION OF RESERVOIR-ASSOCIATED FORESHOCKS AND AFTERSHOCKS

The regularity in the time distribution of foreshocks and aftershocks is a very interesting and important characteristic. Utsu (1961) found that the time distribution of aftershocks is given by the inverse power law:

$$n(t) = Ct^{-h}$$

where $n(t)$ is the frequency of aftershocks per unit time, C and h are con-

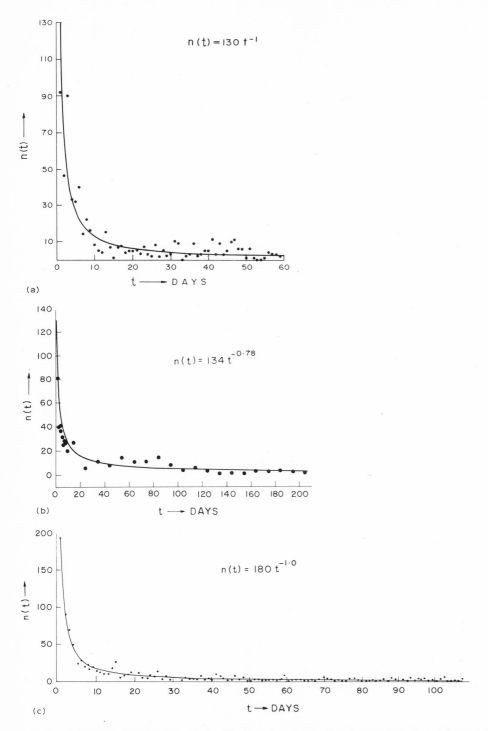

Fig. 87. (a) Time distribution of the Kariba aftershocks for the earthquake of September 23, 1963. (b) Time distribution of the Kremasta aftershocks for the earthquake of February 5, 1966 (after Comninakis et al., 1968). (c) Time distribution of the Koyna aftershocks for the earthquake of December 10, 1967.

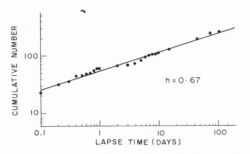

Fig. 88. Cumulative number of aftershocks for the Kurobe earthquake of November 16, 1968 (after Hagiwara and Ohtake, 1972).

stants, and t is the time elapsed after the mainshock. The value of h indicates the rate of decay of aftershock frequency and can be used to infer the physical and stress state of the aftershock zone (Mogi, 1962a). According to Utsu, the frequency curve of aftershocks in the earlier stage is satisfactorily expressed by the above relation up to about 100 days. In a later stage it is well approximated, for many cases, by the following relation:

$$n(t) = C_1 e^{-pt}$$

where C_1 and p are constants. However, Nur and Booker (1972) found that for the 1966 Parkfield—Cholame earthquake sequence, which is one of the best sequences ever recorded, the frequency of aftershocks showed an initial decay of $t^{-1/2}$ to $t^{-3/4}$, gradually shifting to t^{-1} at a later time.

The relation $n(t) = Ct^{-h}$ has been found to hold good for the aftershock sequences at Kariba, Kremasta, Koyna and Kurobe (Figs. 87 and 88). The

TABLE XIII

The relations showing the decay of aftershock activity

Region	$n(t) = Ct^{-h}$	Unit of time	Total time	Source
Kariba	$130t^{-1.0}$	1 day	60 days	Gupta et al. (1972b)
Kremasta	$134t^{-0.78}$	1 day	200 days	Comninakis et al. (1968)
Koyna	$180t^{-1.0}$	1 day	110 days	Gupta et al. (1972b)
Koyna	$1,342t^{-0.77}$	15 days	December, 1967, to December, 1971	Guha et al. (1974)
Kurobe	$Ct^{-0.67}$	cumulative	November 16, 1968, to April, 1970	Hagiwara and Ohtake (1972)

relations in Table XIII give the number of aftershocks on the tth day after the mainshock in these sequences. For the Kurobe aftershock sequence, Hagiwara and Ohtake (1972) have plotted the cumulative number of after-shocks (Fig. 88), however, they did not obtain the value of the constant C.

The values of h are relatively lower for reservoir-associated aftershock sequences and the aftershock activity decayed at a slow pace.

The foreshock distribution of reservoir-associated earthquake sequences have been similarly studied. Guha et al. (1974) used the relation $n(t) = n_1 b^{-t}$ to describe the foreshocks of the Koyna sequence. Here, t is the time interval before the occurrence of the main earthquake. They obtained the following relation for the Koyna foreshocks:

$$n(t) = 8.81 \ (0.0053)^{-t}$$

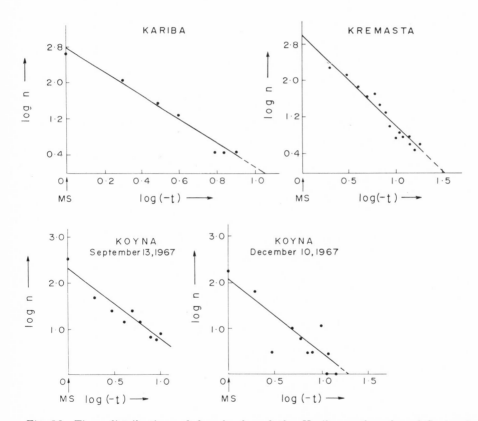

Fig. 89. Time distribution of foreshocks of the Kariba earthquake of September 23, 1963, Kremasta earthquake of February 5, 1966, and Koyna earthquakes of September 13 and December 10, 1967 (redrawn from Papazachos, 1973). The unit of time is five days for Kremasta, one week for Koyna and one month for Kariba.

Papazachos (1973) has shown that the occurrence of the foreshocks could be described by the relation:

$$n(t)\,dt = n_0(\tau - t)^{-h}\,dt\,,\quad t \leqslant \tau_0$$

where t is the time in days, $n(t)dt$ is the number of foreshocks per day, and n_0, τ and h are constants. The theoretical duration of the foreshock sequence, τ_0, is given by $n_0 = (\tau_0 + 1)^h$. The following relations, which give the number of foreshocks on the tth day after the start of the activity, have been obtained for the Koyna, Kariba and Kremasta sequences.

Kariba:

$$n(t) = 125{,}380\,(335 - t)^{-2.60}\,,\quad t \leqslant 305 \text{ days}$$

Kremasta:

$$n(t) = \quad 4{,}736\,(159 - t)^{-1.99}\,,\quad t \leqslant 105 \text{ days}$$

Koyna (September 13, 1967):

$$n(t) = \quad 364\,(124 - t)^{-1.63}\,,\quad t \leqslant 117 \text{ days}$$

Koyna (December 10, 1967):

$$n(t) = \quad 537\,(216 - t)^{-1.53}\,,\quad t \leqslant 209 \text{ days}$$

These relations are shown in Fig. 89.

FORESHOCK—AFTERSHOCK PATTERNS

In the last section we have seen the time distribution of the foreshocks and aftershocks of some of the reservoir-associated earthquake sequences. When the frequency distribution per unit time of both the foreshocks and aftershocks are combined, the pattern thus obtained is called the foreshock—aftershock pattern.

Mogi (1963b) classified the foreshock—aftershock patterns found in the experimental models into three types as shown in Fig. 90, and compared the patterns with those of natural earthquake sequences. The difference between the three types is due to the structural states of materials and the space distribution of the applied stress as described below.

Type 1. In the case of homogeneous material and uniformly applied stress, a mainshock occurs without any foreshock and is followed by numerous elastic aftershocks.

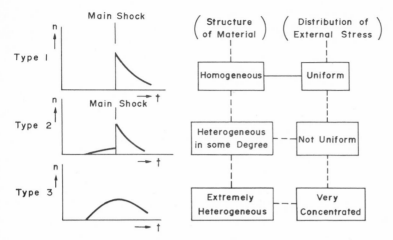

Fig. 90. Mogi's (1963b) three types of foreshock—aftershock patterns and their relationship to the structures of materials and applied stresses.

Type 2. When the material has a rather heterogeneous structure and/or the applied stress is not uniform, small elastic shocks occur prior to a mainshock and many aftershocks occur following the mainshock.

Type 3. When the structure of the material is extremely heterogeneous and/or the applied stress has a considerable concentration, a swarm type of activity occurs when the number of elastic shocks and their magnitude increases gradually and then decreases after some time.

These three typical foreshock—aftershock patterns are also exhibited by natural earthquake sequences. From this comparison, Mogi inferred that the mechanical structures of the media and the nature of the applied stresses must be responsible for these three different foreshock—aftershock patterns of natural earthquakes. Since the stresses could be regarded as nearly uniform for tectonic earthquakes, the pattern of earthquake sequences would be mostly influenced by the degree of heterogeneity of the media (Mogi, 1963b).

It has been shown earlier that the mainshocks at Kariba, Kremasta, and Koyna were preceded by many foreshocks and followed by a large number of aftershocks. Hence the foreshock—aftershock patterns for these earthquakes correspond to Type 2 of Mogi's classification. Fig. 91 shows the foreshock—aftershock pattern for the main Kariba earthquake at 09:01:57 on September 23, 1963. This shock was preceded by about 20 shocks in a period of 24 hours. The frequency of the shocks abruptly reached a maximum at the time of the mainshock. This was followed by a number of aftershocks. In the week following the main earthquake about 360 earthquakes were recorded. The unit of time in this figure is 1 day.

Fig. 92 shows the foreshock—aftershock pattern of the Kremasta earth-

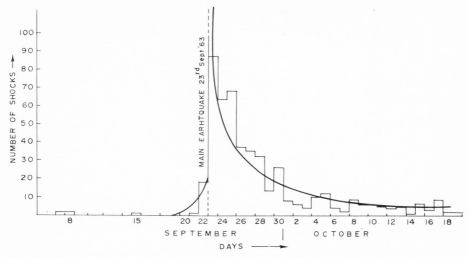

Fig. 91. Foreshock—aftershock pattern for the main Kariba earthquake on September 23, 1963.

quake sequence. The unit of time taken in this figure is 5 days. The earthquakes of magnitude ⩾3.4 listed by Comninakis et al. (1968) have been used for plotting the foreshock—aftershock pattern. The mainshock occurred on February 5, 1966. As seen in the figure, the number of foreshocks increased steadily until the time of the mainshock. The main earthquake was followed by a number of aftershocks. Papazachos et al. (1967) listed 60 large earthquake sequences in the Greece region, of which the mainshocks have magnitudes exceeding 6. Unlike the Kremasta earthquake of February 5, 1966, which was preceded by 17 foreshocks, only a few of these earthquakes were

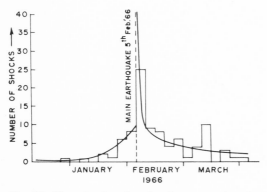

Fig. 92. Foreshock—aftershock pattern for the main Kremasta earthquake on February 5, 1966.

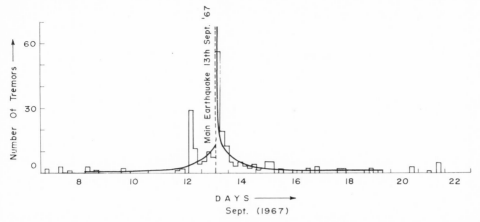

Fig. 93. Foreshock—aftershock pattern for the Koyna earthquake of September 13, 1967.

preceded by a small number of foreshocks. Hence, these earthquakes could be classified under Type 1 of Mogi's classification.

Figs. 93 and 94 show the foreshock—aftershock patterns for the two Koyna earthquakes which occurred on September 13 and December 10, 1967, respectively. Both patterns are identical to Type 2 of Mogi's models. The December 10, 1967, earthquake was preceded by about 90 foreshocks which occurred within a period of 10 days and it was followed by more than 400 aftershocks within a period of 20 days. The frequency of foreshocks increased steadily until the occurrence of the main earthquake. It is interest-

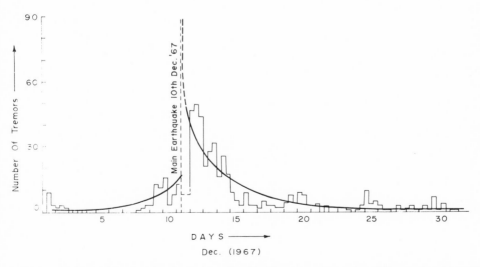

Fig. 94. Foreshock—aftershock pattern for the Koyna earthquake of December 10, 1967.

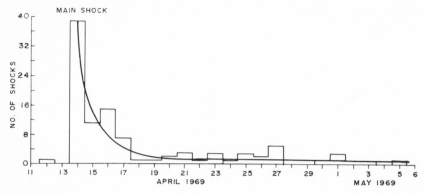

Fig. 95. Foreshock—aftershock pattern for the Godavari Valley earthquake of April 13, 1969.

ing to compare the Koyna earthquake sequence with two recent and compa-rable-magnitude earthquake sequences of Peninsular India, viz. the Godavari Valley sequence of April, 1969, and the Broach earthquake sequence of March, 1970. The Godavari Valley earthquake sequence, whose main earth-quake occurred on April 13, 1969, was fairly well recorded by the N.G.R.I. seismic station at Hyderabad, situated at a distance of about 200 km of the epicenter. For this sequence, earthquakes of magnitudes as low as 2 could be identified from the records of the Hyderabad station. As seen in Fig. 95, the main earthquake was preceded by only one foreshock, whereas the number of aftershocks was quite large. This is a typical example of Mogi's Type 1 model. The Broach sequence, whose main earthquake occurred on March 23, 1970, could not be properly recorded in the absence of any seismograph station in the vicinity. The Hyderabad station recorded 6 aftershocks of magnitude ⩾ 3.2, within the 8 hours following the main earthquake, but no foreshock preceding it. Two aftershocks were recorded on April 26 and August 30, 1970. In the absence of any foreshock activity, the Broach earthquake sequence also falls under Type 1 of Mogi's models.

From the above it can be seen that the normal earthquakes occurring in the Kremasta and Koyna regions belong to Type 1 of Mogi's models, whereas those associated with the reservoirs in these regions belong to Type 2. It is probable that the foreshock—aftershock patterns are influenced by the reser-voirs which affect the structural and/or stress state of the material in the earthquake zone.

FOCAL-MECHANISM SOLUTIONS OF EARTHQUAKES

Focal-mechanism solutions are very useful in determining the orientations of fault planes and stresses. The study of focal-mechanism solution of earthquakes is based upon representing the elastic radiation at large distances as caused by a system of forces acting at the focus of the earthquake. The double-couple or Type 2 model of Honda's (1957) force system is common-ly accepted as representative of the force system at the focus. The initial motion of the P wave from this force system forms a quadrantal distribution of the compressions and dilatations (Nakano, 1923). The planes dividing the quadrants are called the nodal planes, one of which is the fault plane and the other auxiliary plane.

As the ray paths are curved, the observations on the earth's surface are transformed to the surface of a hypothetical unit focal sphere, assuming the focus at its center. Most commonly, the observations of the first motions and the polarization of S waves are plotted on an equal-area projection or Schmidt net (Friedman, 1964). The projection of the lower hemisphere of the focal sphere on the equatorial plane is used. The nodal planes project as median circles. The angles of incidence at the focus of the rays reaching the station are obtained. The azimuth and the angle of incidence are utilized as the two coordinates to plot the observation station on the projection. Tables of the angles of incidence for P and S waves have been published by Nuttli (1969) and Chandra (1970).

In the double-couple source model, the axes that bisect the quadrants of dilatations and compressions are the pressure axis, P, and the tension axis, T, respectively. Intersection of the two nodal planes gives the null axis, B.

The angle of polarization of the S wave is the angle between the plane of polarization and the vertical plane containing the ray and is defined by the relation:

$$\epsilon = \text{Arctan}(SH/SV)$$

In practice, the angle $\bar{\epsilon}$ is obtained by plotting the particle motion of the S phase, that is, by a vector combination of the horizontal components of the S motion. This angle is given by:

$$\bar{\epsilon} = \text{Arctan } \overline{SH}/\overline{SV}$$

where $\overline{SH}$ and $\overline{SV}$ are the horizontal components of SH and SV motion at the free surface. The angle ϵ is obtained from $\bar{\epsilon}$ by the relation:

$$\tan \epsilon = \tan \bar{\epsilon} \cos i_0$$

where i_0 is the angle of incidence of the S wave at the free surface.

The focal-mechanism solutions obtained for the Koyna, Kariba, Kremasta, Lake Mead, Denver, and Rangely earthquakes are described below.

Lake Mead

Rogers and Gallanthine (1974) have determined the focal mechanisms for about 40 microearthquakes occurring during 1972 and 1973 at Lake Mead (Fig. 96). The first five solutions (A—E) represent composite focal mechanisms of five groups of events. In these solutions, the N-striking plane is parallel to the faults in the area. Assuming this plane to be the fault plane, the sense of motion is right-lateral strike-slip. The sense of motion is not in agreement with the inferred motion on the Hamblin Bay and related faults (Fig. 53); however, it agrees with the sense of motion on the Las Vegas shear

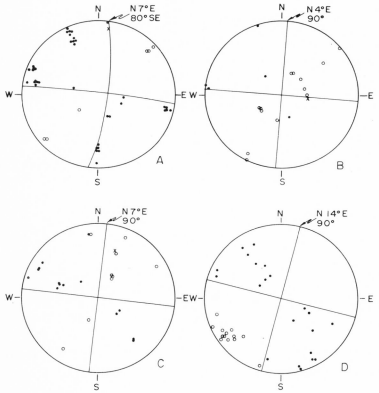

• COMPRESSION
∘ DILATATION
ˣ COMPRESSION AND DILATATION

Fig. 96. Composite focal mechanisms for five groups (A—E) of Lake Mead microearthquakes (after Rogers and Gallanthine, 1974). F and G are the focal mechanisms of two events which did not fit in the composite solutions. Solutions E, F, and G are not uniquely determined due to lack of data, hence the alternate solutions are shown by dotted lines.

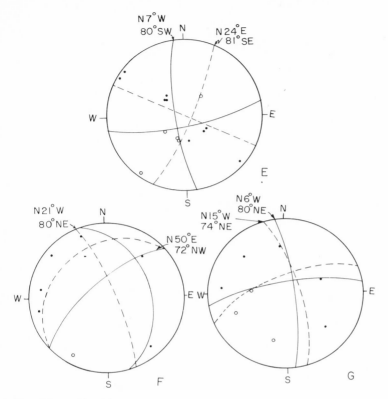

Fig. 96 continued.

zone, which is believed to pass just north of the Boulder Basin. From this observation it has been inferred that currently a right-lateral motion is taking place on the Hamblin Bay shear zone and/or that movement is taking place along some hidden faults related to the Las Vegas shear zone.

Solutions F and G are for the two events which did not fit with the first five solutions. In the solution, shown by dotted lines in Fig. 96F, the sense of motion along the NNW-striking plane is in agreement with the previous solutions, although it has a significant dip-slip component. The dip component indicates that the east side has gone down relative to the west side. Solution G is similar, although it has a lesser dip-slip component than that of solution F. As the events are located near the western margin of the lake (F: 36°2.23′N, 114°46.05′W; G: 36°2.90′N, 114°48.09′W), the lake-side block has gone down (Fig. 58).

Koyna

The different fault-plane solutions reported for the December 10, 1967

TABLE XIV

Different focal-mechanism solutions of the Koyna earthquake of December 10, 1967

Reference	Plane a (deg.)				Plane b (deg.)				Motion with plane a as the fault plane
	strike dir.	dip dir.	dip	slip angle	strike dir.	dip dir.	dip	slip angle	
(1) Gupta et al. (1969)	328	—	90	90	—	—	0	0	normal dip-slip
(2) Tandon and Chaudhury (1968)	26	296	66	15	110	200	74	25	
(3) Lee and Raleigh (1969)	37	307	72	18	125	216	84	4	strike-slip left-lateral for plane a, right-lateral for plane b
(4) Khattri (1970)	350	260	80	11	76	166	70	20	
(5) Sykes (1970)	21	291	75	0	111	201	90	15	
(6) Banghar (1972)	22	292	80	2	112	202	88	10	
(7) Singh et al. (1975)	10	280	78	5	111	21	85	12	
(8) Tsai and Aki (1971)	23	293	70	ca. 15	not given				left-lateral strike-slip

earthquake are given in Table XIV. Solution (1) gives normal dip-slip faulting along a vertical fault plane striking NNW. The northeastern side has moved down relative to the southwestern side. The epicenter of the earthquake was determined to be near the dam; the lake is on the northeastern side of the fault, which is the downthrown side.

To find out this solution, the first-motion data from four African stations were omitted considering their anomalously high "observed minus calculated" residuals of arrival time. However, E.B. Rodrigues (personal communication, 1969) has reported that this may be due to the fact that seismic waves crossing the Eastern Rift system of Africa are slowed down. If these observations are considered as well, a strike-slip solution can be obtained.

In fact, all the other solutions (Table XIV) consistently indicate strike-slip faulting. Those solutions are in close agreement with each other. Singh et al. (1975) have also inferred strike-slip faulting on a near-vertical fault plane which strikes in N10°E direction, using Rayleigh-wave radiation pattern and S-wave polarization angles. Considering the N-striking plane as the fault plane, these solutions indicate a left-lateral strike-slip motion. The lake-side block has moved towards the north or northeast and been downthrown with a small dip component.

Kariba

A fault-plane solution for the Kariba earthquake of September 23, 1963, is shown in Fig. 97. The P-wave first-motion directions and S-wave polarization angles obtained from long-period records are shown in this figure. This fault-plane solution is consistent with the solution of Sykes (1967) given in Table XV. The motion is predominantly dip-slip in both planes associated with normal gravity faulting. If plane *a* is considered to be the fault plane, the small strike-slip component of motion is right-lateral; if plane *b* is considered to be the fault plane then the motion is left-lateral. On the basis of the distribution of epicenters and the NE trend of the faults and fractures near

TABLE XV

Fault-plane solution for the Kariba earthquake of September 23, 1963

Strike direction	Dip direction	Dip	Slip angle
Plane a (deg.)			
0	90	32	70
Plane b (deg.)			
25	294	60	78

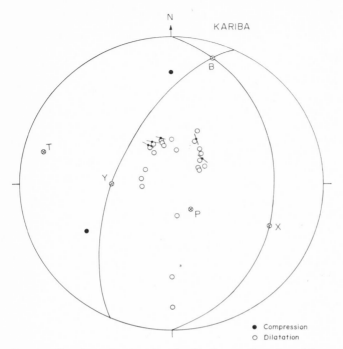

Fig. 97. Focal mechanism for the Kariba earthquake of September 23, 1963.

Kariba Lake (Fig. 45), plane b, striking NNE, could be accepted as the fault plane. The epicenter of this earthquake is near the dam. If plane b is the fault plane, the northwestern side block, on which the lake is situated, has gone down (Gupta et al., 1972b).

Kremasta

For the Kremasta earthquake of February 5, 1966, Comninakis et al. (1968) obtained a fault-plane solution which is shown by broken lines in Fig. 98. The steeply dipping plane, striking WNW, which is cutting through the central field of dilatations in this figure, was accepted by them as the fault plane. The faulting is reverse and sinistrally associated with a near-vertical thrust fault.

The fault-plane solution obtained using additional microfilm data and data which have been reported in the *Bulletin of the International Seismological Center* is shown as a continuous line in Fig. 98 and is given in Table XVI (Gupta et al., 1972b). Five compressions and 19 dilatations observed on the microfilms are plotted as bigger circles and the rest as smaller circles in this figure. The faulting is normal, with the dip component of motion slightly greater than the strike component. The two nodal planes strike ENE (plane a)

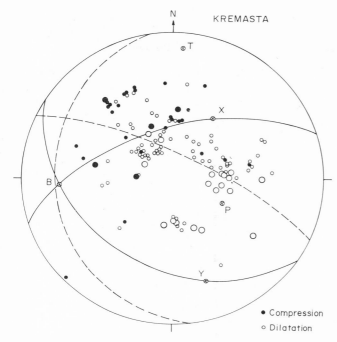

Fig. 98. Focal mechanism for the Kremasta earthquake of February 5, 1966.

and NW (plane *b*). The strike-slip component of the motion is right-lateral along plane *a* and left-lateral along plane *b*.

The tectonics of the Kremasta area are complex, as discussed in Chapter 3. In the context of plate tectonics, subduction of the African plate took place along the Aegean trench from Cretaceous to Miocene times. The thrust faults of the Pindic Orogeny and some other structures may be related to this process. The African plate is now being consumed along the Ionian trench

TABLE XVI

Focal-mechanism solution of the Kremasta earthquake of February 5, 1966

Strike direction	Dip direction	Dip	Slip angle
Plane a (deg.)			
72	342	62	60
Plane b (deg.)			
122	213	40	48

(Dewey and Bird, 1970, p. 2642) and the region is dominated by an E—W compression. The NNW faults belonging to the Hellenide trend (e.g. the Inachos) show the characteristics of thrust faults. Earthquakes associated with these faults are reported to be shallow (A.C. Galanopoulos, personal communication to D.T. Snow, 1971). The Kremasta area lies between the ENE-trending Anatoli and Pec-Scutari transform faults. The Anatoli transform fault is believed to pass through the Gulf of Patrai about 50 km south of the Kremasta Dam. Galanopoulos (1967b) inferred a conjugate set of wrench faults about 10 km north of the Kremasta Dam. Snow (1972) inferred that the present seismicity, in which the mainshock and a number of other shocks were not shallow, is associated with ENE-trending wrench faults, as the Anatoli trend shocks are deep compared to Hellenide trend shocks. He explained that shallow thrust faults may coexist with wrench faults, as weak, horizontal, sliding surfaces of halite rocks are present in the area.

Near the Kremasta Dam, several steep normal faults trending NW have been traced. The main Kremasta earthquake has been located by Comninakis et al. (1968) to be about 25 km north of the dam near the Pindus thrust, which is not active now. No other fault has been traced near this location. However, about 10 km north of the dam both ENE- and WNW-trending faults have been traced. Along the ENE-trending, Alevrada—Smardacha Fault (Fig. 49), equal components of dip-slip and right-lateral strike-slip movement have been observed; this movement is very similar to the inferred nature of motion from the fault-plane solution. Several other ENE-trending steep faults have been mapped which have dextrally displaced the axes of Miocene folds. These are probably the youngest faults of the area. Terra Consult (1965, as quoted in Snow, 1972) have mapped a closeby WNW fault near Trichlinon (Fig. 49).

Considering either of the nodal planes as the fault plane, the dip component of motion on both of these planes is such that a downthrown movement for the lake-side block is indicated. Drakopoulos (1973) has also mentioned that his fault-plane solutions of the main earthquake and a large aftershock of the Kremasta area indicate the dip-slip component of the motion to be such that the lake is situated on the downthrown block.

Denver

Healy et al. (1968) have published fault-plane solutions for some Denver events (Fig. 99). Most of the earthquakes indicate a strike-slip motion on the vertical planes striking either W to NNW or N to ENE. The epicentral zone trends in the WNW direction and is parallel to the first set of nodal planes. Hence the planes striking W to NNW are taken as the fault planes along which the sense of motion is right-lateral.

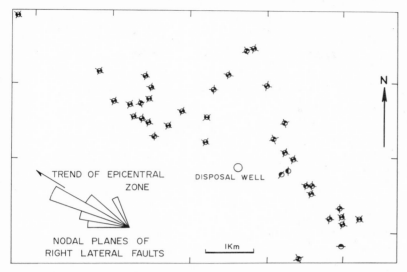

Fig. 99. Focal mechanisms for the Denver earthquakes (after Healy et al., 1968).

Rangely

Focal mechanisms have been determined for the Rangely earthquakes by Raleigh et al. (1972) using the P-wave first-motion directions recorded by 10 or more stations. Most of these mechanisms (Fig. 100) indicate a strike-slip movement on the vertical plane trending either NW or NE. The NE-trending nodal planes showing right-lateral movement are parallel to the trend of the epicentral zone and are likely to represent the fault planes. Faulting along

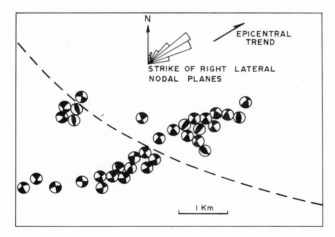

Fig. 100. Focal mechanisms for the Rangely earthquakes (after Raleigh et al., 1972).

these planes is also consistent with the fractures and exposed faults in the area. For these mechanism solutions, the nodal plane and slip-direction orientations vary by $\pm 20°$ for the different earthquakes along different parts of the fault zone.

From the above we have seen that the normal displacement inferred from the fault-plane solutions of the reservoir-associated earthquakes indicates that the lake is situated on the downthrown block. This observation suggests that the fault movement may be due to the vertical load of reservoir water. However, the effect of increased pore pressure due to reservoir filling, which triggers faulting by the reduction in the effective normal stress across a critically stressed fault, looks more convincing (Chapter 6). Both the normal displacements and the strike-slip displacement inferred for the reservoir-associated earthquakes may be triggered in this manner. The strike-slip faulting inferred for the earthquakes in the Denver and Rangely areas is also in accordance with this hypothesis.

INFERRED CHANGES IN THE MECHANICAL PROPERTIES OF CRUSTAL
LAYERS BY RESERVOIR IMPOUNDING

The important findings of the study of the behavior of reservoir-associated earthquakes are:

(1) The foreshock b value is higher than the aftershock b value, both being, in general, higher than the b values for the normal earthquake sequences of the regions concerned and the regional b values.

(2) In addition to a high b value, the magnitude ratio of the largest aftershock to the mainshock is also high.

(3) Aftershocks have a comparatively slow rate of decay.

(4) The foreshock—aftershock patterns are identical and correspond to Type 2 of Mogi's models, whereas the normal earthquake sequences of the regions in question belong to Type 1 of Mogi's models.

All these factors are governed by the mechanical properties of the media; their deviation from normal earthquakes indicates changes in these properties.

As discussed earlier, laboratory experiments have shown that homogeneous media are characterized by earthquakes with no foreshocks, slightly heterogeneous media have a number of foreshocks preceding the main earthquake, and extremely heterogeneous media are characterized by a swarm type of earthquake activity. It has been also shown experimentally that the b value increases with the increase of heterogeneity of the media and is also governed by the ratio of the existing stress within the rock sample to the final breaking stress. The foreshock b values are found to be much lower than the aftershock b values for a natural earthquake sequence; a high b value (> 0.5) in a foreshock sequence is considered by Berg (1968) to be an indication for the occurrence of a high-magnitude earthquake. On the basis of the experimental results of Mogi and Scholz, Berg has explained that, before a main earthquake, the high stress and strength are associated with low b values, whereas high b values for an aftershock sequence correspond to the reduced strength and low stress after the mainshock. The h value (the rate of decay of the aftershocks) for the Japanese region has been found by Mogi (1962a) to range from 0.9 to 1.8, most of the values being greater than 1.0. For reservoir-associated earthquake sequences, the h values are $\leqslant 1$, indicating a slow rate of decay of the aftershocks. For the California earthquake sequences, and other typical earthquake sequences with low b values (0.4—0.5), the largest aftershock to the mainshock magnitude ratio is found to be high (0.9), and with high b values (0.6—0.8) the magnitude ratios are low (0.6—0.7). The earthquake sequences associated with reservoirs are exceptional since both the magnitude ratio and the b values are simultaneously high.

The common features delineated for earthquakes associated with reservoirs, which also differentiate these earthquakes from the normal earthquakes of the regions concerned, and the detailed laboratory studies carried out by various investigators, as mentioned above, suggest that the reservoirs have probably affected the mechanical properties of the media. In a medium having high stress and strength, the increased pore pressure due to reservoir filling initiates the failure of otherwise unfailing fractures and thus introduces heterogeneity. The conceptual model outlined in the following shows how the reduction in competence and the increase in heterogeneity allows the stresses to be relieved from small volumes and brings about the above-mentioned changes in earthquake characteristics.

Let us consider a certain relatively homogeneous rock volume, V, which has been accumulating strain due to certain tectonic processes. As the medium is homogeneous, no foreshocks relieving small amounts of strain would occur. When the accumulated strain exceeds the competence of the rock volume, a main earthquake releasing most of the accumulated strain occurs. This would be followed by some "marginal-adjustment" aftershocks of comparatively much smaller magnitudes. Such an earthquake sequence would belong to Type 1 of Mogi's models and would be associated with a low b value in the frequency—magnitude relation. Also, the ratio of the magnitude of the largest aftershock to the mainshock would be small ($\leqslant 1$). Now, let us see what happens when the mechanical properties of the rock volume, V, change and the medium becomes heterogeneous to some extent at a stage when the accumulated strains have not exceeded the original competence of the rock volume under consideration. The heterogeneity introduced would result in the division of the rock volume, V, into smaller volumes, V_1, V_2, V_3, ... V_n; each volume being capable of releasing its stored energy as and when its competence is exceeded. The seismicity in such a region would be characterized by the occurrence of a number of earthquakes of comparable magnitudes and a high foreshock activity before any major event. Consequently, it would fall under Type 2 of Mogi's models, it would have a higher foreshock b value than that of the aftershock, both being higher than the b value for natural earthquake sequences and the regional b value and, also, the magnitude ratio of the largest aftershock to the mainshock would be high.

On the basis of the above discussion, it may be said that the creation of artificial lakes changes the mechanical properties of the media, making them less competent and hence unable to withhold the accumulated strains. This seems reasonable, considering that the Kariba, Kremasta, and Koyna regions where moderate-magnitude earthquakes ($M \geqslant 6$) occurred are characterized by a volcanic past and by the presence of fractured rocks through which water could permeate (Chapter 3), thus decreasing the competence of initially competent strata. For earthquakes of considerable magnitude to be associated with artificial reservoirs, it seems that initially strained competent strata capable of withholding tectonic strains which later on change their mechanical properties when water is impounded, are vulnerable.

Chapter 5

INCREMENTAL STRESS AND DEPRESSION
DUE TO THE WATER LOAD

In the absence of suitable experiments, facilities and techniques which have been developed in recent years, such as hydro-fracturing, have been conducted at reservoir sites to determine the initial stresses. However, it has been difficult to assess the total stress in the crustal layers near the reservoirs following their impoundment. Moreover, the strengths of the existing faults, i.e. the maximum shear stress which could exist across a fault without causing its failure, are also not known. Another important factor is the change of pore-fluid pressure following reservoir impounding, and its effect on the effective normal stress which is also unknown. However, incremental stresses caused by the load of the reservoir water can be estimated analytically. From these estimates, it is possible to determine the increment of the shear stress along the existing faults, provided the fault-plane orientation is known. It is necessary to mention that the magnitude of the incremental stress caused by reservoir loading seldom exceeds 10 bars, which is small compared to the strength of rocks (crystalline rocks have a strength of about 1,000 bars). However, Gough and Gough (1970b) consider that such increments could also trigger earthquakes in critically stressed fault environments.

Incremental stresses have been calculated for two-dimensional lakes of infinite length (Gough, 1969), as well as for three-dimensional lakes of any shape (Gough and Gough, 1970a). In the following we outline the procedural details for calculating the incremental stresses for the two cases.

STRESS EQUATIONS FOR A TWO-DIMENSIONAL CASE

The two-dimensional lake is approximated by an infinitely long, water-filled trough in the plane surface on an elastic halfspace. The cross-section of the two-dimensional trough can be approximated by a polygon of n sides. Let $ABCD \dots N$ be such a polygonal section (Fig. 101). The line BC represents a plane face of the lake bottom of infinite length perpendicular to the section. Gough (1969) derived the following expressions for the normal stresses σ_u and σ_v, and the shear stress τ_{uv} produced at P due to BC (Fig.

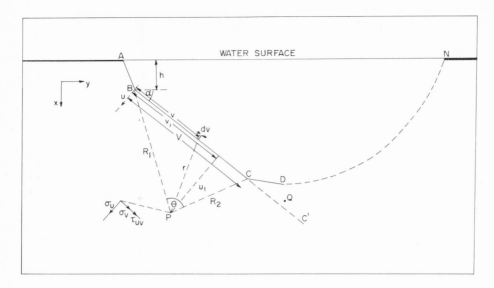

Fig. 101. Polygonal section of a two-dimensional lake with the notations used for computing the incremental stress.

101), taking u and v as the axes, respectively perpendicular and parallel to BC:

$$\sigma_u = \frac{-\rho g u}{\pi} \{(h + v \sin \alpha)[(V - v)R_2^{-2} + v R_1^{-2} + \theta u^{-1}]$$

$$+ u^2 \sin \alpha (R_1^{-2} - R_2^{-2})\}$$

$$\sigma_v = \frac{-\rho g u}{\pi} \{(h + v \sin \alpha)[\theta u^{-1} - (V - v)R_2^{-2} - v R_1^{-2}] \qquad [5.1]$$

$$+ \sin \alpha \ln(R_2^2 R_1^{-2}) + u^2 \sin \alpha (R_2^{-2} - R_1^{-2})\}$$

$$\tau_{uv} = \frac{-\rho g u^2}{\pi} [h(R_1^{-2} - R_2^{-2}) + \sin \alpha(\theta u^{-1} - V R_2^{-2})]$$

where ρ is the density of water, u and v are coordinates of P with respect to B as the origin, V is the width of the strip BC, θ is the angle which it subtends at P, $R_1^2 = u^2 + v^2 = PB^2$, and $R_2^2 = u^2 + (V - v)^2 = PC^2$.

At points along the strip BC, $u = 0$, $\theta = \pi$ and the above equations reduce to:

$$\sigma_u = \sigma_v = -\rho g(h + v \sin \alpha)$$

and:

$$\tau_{uv} = 0$$

At points along the extension of the strip on either side, $u = 0$, $\theta = 0$. All stress components vanish and the strip BC contributes no stress at points located so that $u < 0$, $v < 0$ or $u < 0$, $v > V$. These considerations simplify the computations to a large extent.

The summing-up of the stresses contributed by all faces of the section is conveniently done by rotating the axes to a new coordinate system (x, y) having the axes in vertical and horizontal directions, using the following relations given by Jaeger (1956). The horizontal axis is in the plane of the section:

$$\sigma_x = \sigma_u \cos^2 \alpha + \sigma_v \sin^2 \alpha + 2\tau_{uv} \sin \alpha \cos \alpha$$

$$\sigma_y = \sigma_u \sin^2 \alpha + \sigma_v \cos^2 \alpha - 2\tau_{uv} \sin \alpha \cos \alpha \qquad [5.2]$$

$$\tau_{xy} = \tau_{uv} (\cos^2 \alpha - \sin^2 \alpha) + (\sigma_v - \sigma_u) \sin \alpha \cos \alpha$$

The contributions of the other faces of the section are added to σ_x, σ_y and τ_{xy} to arrive at the stress components at P.

The shear stress, τ_{xy} (the total from all n faces), is that which acts across the vertical and horizontal planes through P. If a fault is known to exist near P having a dip β in the xy-plane, the shear stress across it can be found by the last equation. The maximum shear stress, τ_{max}, at P, and the dips, I_m and $I_m + \pi/2$, of the planes through P across which τ_{max} acts could be found by the following relations (Jaeger, 1956):

$$\tau_{max} = \tfrac{1}{2} [(\sigma_x - \sigma_y)^2 + 4\tau_{xy}^2]^{\tfrac{1}{2}} \qquad [5.3]$$

$$I_m = \tfrac{1}{2} \text{Arctan} \left(\frac{2\tau_{xy}}{(\sigma_x - \sigma_y)} \right) \qquad [5.4]$$

τ_{max} is half the difference between the principal stresses at P, which can be computed. The vertical normal stress, σ_x, has special importance where normal faults exist. The Kariba reservoir is a typical example of a lake which has been impounded in a rift valley. For such cases, the initial principal stresses may be tensional in the horizontal direction and compressional in the vertical direction. Under these circumstances, the vertical compressional stress would be enhanced by σ_x due to the load of the added water.

Five parameters, σ_x, σ_y, τ_{xy}, τ_{max} and I_m, are calculated for a grid of 100 field points in the x-direction and 100 points along the y-direction, giving a total of 10^4 points. The output for the incremental stresses is

obtained in the form of computer plots which give the percentage of maximum values. Such contours of downward normal stress, maximum shear
stress, and the contours of the inclination of planes across which maximum
shear stress acts have been shown by Gough (1969) for two simple models,
lakes of rectangular and isosceles-triangular (Fig. 102) sections, and a 17-
corner polygon.

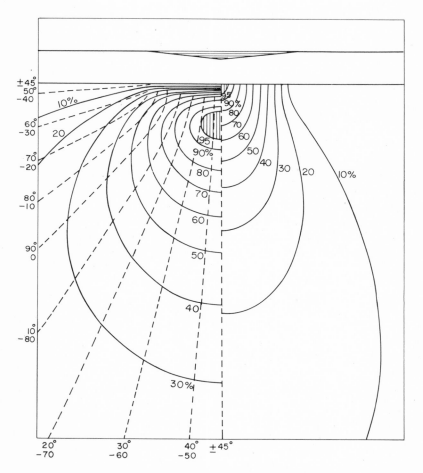

Fig. 102. Incremental stress beneath a lake with a triangular cross-section of 20 km width
and 100 m maximum depth (vertical exaggeration is 10 ×). The downward normal stress
shown in the right half of the figure is given as percentage of the maximum value 9.33
bars. Maximum shear stress on the left is shown as percentage of the maximum value 2.41
bars. Broken lines give inclinations of planes across which maximum shear stress acts
(after Gough, 1969).

STRESS EQUATIONS FOR A THREE-DIMENSIONAL CASE

In the three-dimensional case, the lake is divided into small squares of side a by two sets of orthogonal straight lines, one in the east—west direction and the other in the north—south. The integration is carried out numerically and hence requires an execution time ten times longer than that for the two-dimensional case where integration is done analytically. Moreover, for the three-dimensional case, the assumptions are such that the stresses cannot be calculated at the lake floor since the calculations start at a depth of $4a/3$ from the lake floor. However, the results of these calculations could easily be extrapolated for the lake-floor values.

After estimating the mean water depth, h, in every square, the vertical force $F = \rho g a^2 h$, acting at the center of the square, could be substituted for the water pressure. Thus a two-dimensional array of forces F, acting normally to the boundary plane of an elastic halfspace at spacings a, replaces the lake.

Taking the x-axis eastwards, the y-axis northwards in the plane boundary and the z-axis downwards, the distance of point $P(x,y,z)$ from the origin is given by $R = \sqrt{(x^2 + y^2 + z^2)}$ and its horizontal projection is given by $r = \sqrt{(x^2 + y^2)}$. The following three normal stress components and one shear stress component give the stress at point P:

$$\sigma_r = \frac{F}{2\pi} \left[\frac{1 - 2\nu}{r^2} \left(1 - \frac{z}{R}\right) - \frac{3r^2 z}{R^5} \right]$$

$$\sigma_z = -\frac{3F}{2\pi} \frac{z^3}{R^5}$$

$$\sigma_\theta = \frac{F}{2\pi} (1 - 2\nu) \left(-\frac{1}{r^2} + \frac{z}{r^2 R} + \frac{z}{R^3}\right)$$

$$\tau_{rz} = -\frac{3F}{2\pi} \frac{rz^2}{R^5}$$

[5.5]

where ν is Poisson's ratio, and the azimuthal angle, $\theta = \text{Arctan}(y/x)$, is measured from east through north (Timoshenko and Goodier, 1951, p. 364). For adding the contributions of many forces F, the axes are rotated by means of the relations:

$$\sigma_x = \sigma_\theta \sin^2\theta + \sigma_r \cos^2\theta \qquad \tau_{xy} = (\sigma_r - \sigma_\theta) \sin\theta \cos\theta$$

$$\sigma_y = \sigma_r \sin^2\theta + \sigma_\theta \cos^2\theta \qquad \tau_{xz} = \tau_{rz} \cos\theta$$

$$\sigma_z = \sigma_z \qquad \tau_{yz} = \tau_{rz} \sin\theta$$

[5.6]

The stress as a result of the load of the entire lake at any point is obtained through the summation of the contributions from all forces F to the six stress components, i.e. σ_x, σ_y, σ_z, τ_{xy}, τ_{yz}, and τ_{zx}. These computations are carried out using in turn relations [5.5] and [5.6] for every force F. The principal stresses σ_1, σ_2 and σ_3 at this point are eigenvalues of σ such that:

$$\begin{bmatrix} \sigma_x - \sigma & \tau_{yx} & \tau_{zx} \\ \tau_{xy} & \sigma_y - \sigma & \tau_{zy} \\ \tau_{xz} & \tau_{yz} & \sigma_z - \sigma \end{bmatrix} = 0 \qquad [5.7]$$

and the corresponding eigenvectors are the direction cosines l_1, m_1, n_1 of σ_1, and similarly for σ_2 and σ_3, as discussed by Jaeger (1956, p. 13).

From the various parameters that could be chosen for output, Gough and Gough (1970a) have chosen the downward normal stress σ_z, the maximum shear stress $\tau_{max} = \frac{1}{2}(\sigma_1 - \sigma_3)$, and the orientation of the two orthogonal planes across which τ_{max} acts at this particular point. The azimuths of the downward normal to these two planes are:

$$A_1 = \operatorname{Arctan}\left(\frac{l_1 + l_3}{m_1 + m_3}\right)$$

$$[5.8]$$

$$A_2 = \operatorname{Arctan}\left(\frac{l_1 - l_3}{m_1 - m_3}\right)$$

which are clockwise from the north, i.e. from y through x. The planes are inclined to the horizontal by:

$$I_1 = \operatorname{Arctan}\frac{\sqrt{(l_1 + l_3)^2 + (m_1 + m_3)^2}}{n_1 + n_3}$$

$$[5.9]$$

$$I_2 = \operatorname{Arctan}\frac{\sqrt{(l_1 - l_3)^2 + (m_1 - m_3)^2}}{n_1 - n_3}$$

THE INCREMENTAL STRESS DUE TO THE WATER LOAD OF LAKE KARIBA

For Lake Kariba, calculations of the incremental stress were made assuming a three-dimensional case. As the lake has an E—W trend where it is deep, the field points have been chosen in 26 N—S vertical (yz) planes, each of which gives the parameters σ_z, τ_{max}, A_1, A_2, I_1 and I_2 over a vertical section (Gough and Gough, 1970a). The lake has been divided into 1,302

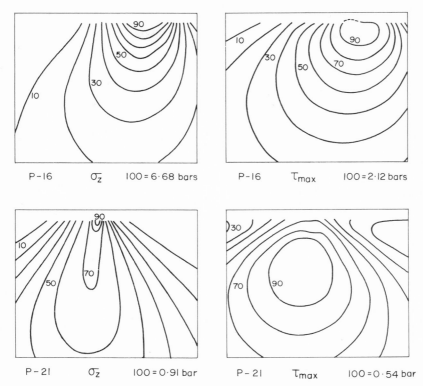

Fig. 103. Downward normal stress (σ_z) and maximum shear stress (τ_{max}) in two N–S vertical planes (*P-16* and *P-21*) across Lake Kariba, north being at the right. The frames are 45 km wide and 37.5 km deep. Parameters assumed are 0.27 for Poisson's ratio and 0.85 Mbar for Young's modulus (after Gough and Gough, 1970a).

nearly square rectangles of 2.22 km × 2.30 km. An array of n_y = 51 and n_z = 24 field points for each section has been taken. The output for σ_z and τ_{max} has been shown in the form of the percentage of maximum values (Fig. 103). The largest value of τ_{max} for the lake is 2.12 bars, which occurs at about 5 km depth under the Sanyati Basin (Fig. 47b), and the maximum downward normal stress is 6.68 bars.

Gough and Gough (1970a) found that the τ_{max}-planes strike NE–SW to E–W (Fig. 104) and that they are roughly parallel to the strike of the faults shown in Fig. 45. In the Sanyati Basin, where most of the epicenters are located, the τ_{max}-planes strike about N80°E and the faults about N50°E. Here, the shear stress across a fault can not exceed $(\sqrt{3/2})\tau_{max}$. The shear stress across a fault plane at P would be τ where $0 \leqslant \tau \leqslant \tau_{max} \cos \theta$, θ being the angle between the two above-mentioned strike directions. Gough and Gough (1970b) have used the calculations of the stress distributions at nine lake levels for estimating the volume V_τ within which the maximum shear

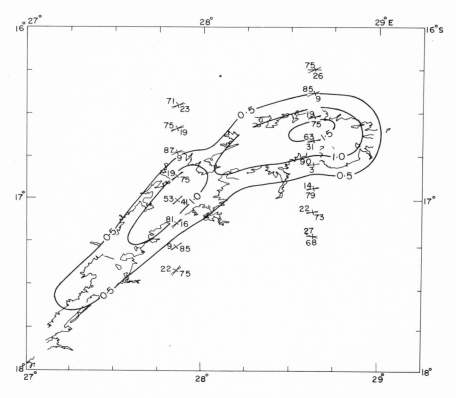

Fig. 104. Maximum shear stress contours at a depth of 13 km under Lake Kariba and the strike and dip of pairs of planes across which maximum shear stress acts.

stress exceeds 1 bar; V_τ as a function of time has been compared with the seismic activity. The increase in seismic activity showed a good correlation with the rise of V_τ until mid-1966, but not thereafter.

Modes of failure

The tremors in the Kariba region could be divided into three groups:

(1) Tremors near the Binga Fault (Fig. 47a).

(2) Tremors occurring downstream of the Kariba Dam, in the rift valley between Kariba and Chirundu (Fig. 47b).

(3) Tremors in and near the Sanyati Basin where most of the activity, including all the large events, occurred (Fig. 47b).

The Binga Fault is suspected to have been active before the filling of the lake. In the initial stages of filling, during 1961—1962, seismic activity was mainly confined to this fault. No tremors were located here in 1963. But there is no certainty that the Binga Fault was completely inactive in 1963, as the seismograph which operated at Binga was shifted to Kamativi (Fig. 47a). The peak seismic activity on this fault, which accounted for the majority of earthquakes, occurred in June, 1961. By this time, the maximum shear stress added by the lake at the Binga Fault would have been about 0.4 bar. As the fault was submerged under water, an increase of pore pressure must have also occurred. Hence, either the incremental stress or the increased pore pressure, or both, might have activated the initially stressed fault.

During 1961, a few tremors were located as far as Chirundu, which is about 50 km north of Kariba. As the distance of these epicenters from the lake was quite large and there was no indication of a raised water table in the area, Gough and Gough (1970b) argued that the lake could not have increased the pore pressure at these distances. According to them, the added shear stress was 0.06 bar at the farthest point and 0.3 bar at the nearest point from the lake, and this additional shear stress probably triggered the critically stressed faults. However, as discussed in Chapter 6, Snow (1974) has shown that the pore pressure in this area may have increased.

For the seismic activity of the Sanyati Basin, Gough and Gough (1970b) considered three modes of failure: (1) the direct effect of the added shear stress, (2) an added stress triggering the failure of the initially stressed faults, and (3) the effect of increased pore pressure. They believe that the increase in pore pressure might have had only a minor effect on the seismicity of the Sanyati Basin. As an increase in the volume V_r with time correlates with enhanced seismic activity, Gough and Gough (1970b) suggested that the shear stress added by the lake caused failure, either directly or indirectly, by triggering the critically stressed faults. However, due to the following considerations, faults cannot fail only as a result of the very small shear stress which are added by the lake.

The shear strength of a dry fault (devoid of pore pressure) is given by:

$$\tau_s = \mu g \rho Z \left(1 - \frac{1 - 2\nu}{1 - \nu} \sin^2 \alpha \right) \tag{5.10}$$

where ν is Poisson's ratio, Z is the depth, ρ is the density of the medium, μ is the coefficient of friction, and α is the inclination of the fault plane with respect to the horizon. τ_s is the least for vertical faults, with a minimum value of:

$$\tau_{sm} = \frac{\mu g \rho Z \nu}{(1 - \nu)} \tag{5.11}$$

Since Krsmanovic (1967) has shown that $\mu = 0.65$ for sandstone and 0.70 for conglomerate, Gough and Gough (1970b) assumed that $\mu = 0.5$—1.0 for gneiss and other basement rocks beneath Lake Kariba. For a dry, vertical fault with $\mu = 1$ and with $\nu = 0.27$, $g = 980$ cm/sec^2 and $\rho = 2.7$ g/cm^3, τ_{sm}

= 2 bars at a depth of 20 m; if μ = 0.5, this strength is reached at 40 m. According to Gough and Gough (1970b), a shear stress of this order cannot activate a dry fault. If water is present at the hydrostatic pressure $p = g\rho_w Z$, and if the rock is permeable so that the water pressure is transmitted through it, the minimum shear strength for a vertical fault would be τ_{sm} = $\mu g(\rho - \rho_w)Zv/(1 - v)$ and would reach 2 bars at a depth of 64 m if μ = 0.5.

Hence, faults, even if wet, cannot be activated only as a result of a shear stress of the order of 2 bars caused by the weight of the lake water. Gough and Gough (1970b) argued that the incremental shear stress due to the water load has triggered failure of faults which were initially stressed at several tens of bars. The added normal stress increases the vertical compression in the normal fault environment, thereby increasing the stress difference between the vertical compression and horizontal tension, which could trigger failure (Chapter 6).

THE INCREMENTAL STRESS UNDER OTHER RESERVOIRS

A reservoir is being formed over the Zambezi River downstream of Kariba at Cabora Bassa in Mozambique. Gough and Gough (1973) have calculated the incremental stress caused by the water load on the known faults in the gorge on which the dam has been constructed. The load near the dam has been treated as a two-dimensional case, and the shallow reservoir upstream of the dam has been treated as a three-dimensional case. They have developed a program combining the algorithms for the dam and the reservoir.

For the Koyna reservoir Guha et al. (1974) showed that the maximum shear stress occurs at about 5 km depth. They plotted the contours of 70% and 50% of maximum shear stress along an E—W section near the dam (Fig. 43b).

THE DOWNWARD ELASTIC DEFLECTION DUE TO THE WATER LOAD OF LAKE KARIBA

The vertical deflection Δd at the point P as a result of the force F is given by:

$$\Delta d = \frac{F}{2\pi E}\left[\frac{(1 + v)Z^2}{R^3} + \frac{2(1 - v^2)}{R}\right] \qquad [5.12]$$

where E is Young's modulus and R is the distance of point P from the origin (Timoshenko and Goodier, 1951). The deflections due to all the point forces are added to give the deflection d at P as a result of the water load of the lake.

Earlier we have discussed the incremental stresses calculated by Gough and Gough (1970a) along 26 sections of Lake Kariba. They also calculated the deflection using relation [5.12] along these sections. They combined the data from all the sections in the form of a contour map of the vertical deflections along a horizontal plane at a depth of 3 km. The maximum deflection was found to be 23.5 cm in the Sanyati Basin.

Before the formation of the lake, extensive precise levelling was carried out in the area in 1953 and re-levelling of some of the routes was done in 1968 (Sleigh et al., 1969) which provides information about the deflection. Gough and Gough (1970a) compared the calculated deflections with the observed deflections along a 50-km levelling route, which extends from Kariba to Makuti due NE, and found a close agreement. The calculated and observed deflections are found to ·coincide precisely, being of the order of 12 cm.

THE DEFLECTION CALCULATED BENEATH OTHER RESERVOIRS

Westergaard and Adkins (1934) computed the amount by which Lake Mead Basin sagged as a result of the water load by applying Boussinesq's equation to a large number of uniformly divided areas of the reservoir basin. The calculated sagging was found to be correct by a triangulation survey of the reservoir basin.

Using Gough's method, Green (1974) calculated the depression under the Hendrik Verwoerd reservoir, South Africa. The maximum depression obtained was 31.7 mm at 1 km depth.

From the foregoing, it is concluded that the incremental stresses due to the water load of impounded reservoirs are rather small. For Lake Kariba, which is one of the largest reservoirs in the world, the computed increment in the vertical stress was about 7 bars and the maximum increment in shear stress was of the order of only 2 bars. For such small stress changes to cause earthquakes, the region has to be critically stressed before impounding. The calculated depression in the vicinity of the reservoir following the impounding is found to tally with the observed depression.

Chapter 6

THE PART PLAYED BY PORE PRESSURES IN INDUCING EARTHQUAKES

The classical example of the earthquake swarm experienced following the injection of waste fluid at high pressure into the Rocky Mountain Arsenal well at Denver, Colorado, has drawn the attention of earth scientists all over the world to the part played by the pore-fluid pressures in causing such earthquakes. Earthquakes have also been clearly linked to the fluid injection in deep wells at the Rangely oil field in northwestern Colorado.

The mechanics of triggering earthquakes by fluid-pressure increases has been provided by Hubbert and Rubey (1959). On the basis of their hypothesis, Evans (1966) attributed the Denver earthquakes to a reduction in the effective normal stress across pre-existing faults in the reservoir rock by the injection of fluid. This view has been further supported through a detailed analysis by Healy et al. (1968). The Rangely oilfield experiments provided the first adequate field test of Hubbert and Rubey's (1959) hypothesis, which is discussed in later sections. The Rangely experiments were so designed that the fluid pressure in the epicentral zone was lowered or increased in order to determine to what extent the fluid pressure controls the earthquake occurrences. It was possible in this case to calculate the distribution of stresses and the fluid pressure in the entire oil field as a number of wells were available for measurements. However, in the case of Denver and other seismic reservoir sites, it has not been possible to measure the stress and fluid-pressure distribution near the seismic areas and to investigate the possible part played by the fluid-pressure variations on the stress distributions.

In the following we discuss the effect of pore pressure on the stress distribution, and Snow's analysis of the changes of effective stresses caused by pore-pressure variations in different fault environments and reservoir geometry. The seismicity at the Denver and Rangely injection wells, and the artificial lakes at Kariba, Kremasta and Koyna (where earthquakes of magnitude exceeding 6 have occurred) have been comprehended as due to pore-pressure changes. The latest developments in experimental and theoretical model studies for calculating the changes in effective stress caused by pore pressure are briefly mentioned.

STRESS RELATIONS

The normal stress σ and the shear stress τ across a plane in an elastic medium are given by:

$$\tau = \frac{\sigma_1 - \sigma_3}{2} \sin 2\alpha \qquad\qquad [6.1]$$

$$\sigma = \frac{\sigma_1 + \sigma_3}{2} + \frac{\sigma_1 - \sigma_3}{2} \cos 2\alpha \qquad\qquad [6.2]$$

where σ_1 and σ_3 are the greatest and the least-principal stresses (compressive stresses being taken as positive), and α is the angle that the plane makes with the σ_3-axis (Jaeger, 1956). The Mohr diagram is the simplest and most useful way of graphically representing the stress relationship. The normal stress and the shear stress are chosen as the abscissa and the ordinate. From the mid-point of σ_3 and σ_1, a radius vector of length $(\sigma_1 - \sigma_3)/2$, at an anticlock-wise angle 2α with the positive direction of the σ-axis, is erected (Fig. 105). The values of σ and τ satisfying [6.1] and [6.2] for certain fixed values of σ_1 and σ_3 for different values of α would lie on this circle which passes through the points σ_3 and σ_1, known as the Mohr circle. It provides the values of the normal and shear stresses across the plane of reference for any angle α and any combination of the values σ_3 and σ_1.

The results of triaxial tests, where for a certain confining pressure applied radially (σ_3), the axial stress (σ_1) is increased until failure occurs, are con-

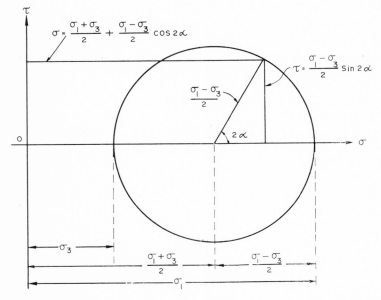

Fig. 105. Representation of normal and shear stresses produced by the greatest and least-principal stresses on the Mohr diagram.

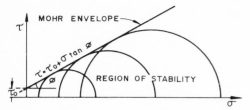

Fig. 106. Mohr envelope of stress circles showing failure for various values of confining pressures.

veniently represented by the Mohr diagram. For a given value of σ_3 and a corresponding value of σ_1, before failure, there is no combination of τ and σ in the specimen which has reached the strength of the material. When finally failure occurs, a combination must have been reached; however, it can not be known from a single test made from the infinite number of possible pairs. When a series of similar tests is made with varying values of σ_3, the envelope of the successively overlapping circles must be the loci of these critical values, since the Mohr circle becomes tangent to this envelope when the fracture occurs. The loci of these critical values is approximated by the straight line (Fig. 106):

$$\tau = \tau_0 + \sigma \tan \phi \qquad \qquad [6.3]$$

This was anticipated by Coulomb (1776) and hence is known as Coulomb's law of failure. τ_0 is the initial shear strength of the rock when the normal stress σ is zero. The angle ϕ is an intrinsic property of the rock and is known as the angle of internal friction, and:

$$\tan \phi = \frac{\tau - \tau_0}{\sigma} \qquad \qquad [6.4]$$

is known as the coefficient of internal friction.

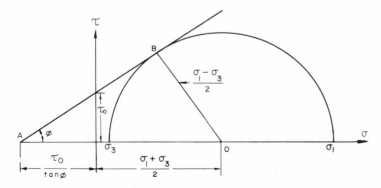

Fig. 107. Relationship between σ, τ and σ_1, σ_3.

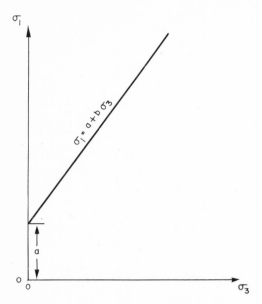

Fig. 108. Failure according to Coulomb's law represented in terms of σ_1 and σ_3.

If the Mohr envelope is approximated by a pair of straight lines in accordance with Coulomb's law [6.3], then $\angle ABO$ would be $\pi/2$ and the following relation (Fig. 107) between σ_3 and σ_1 would hold good:

$$\frac{\sigma_1 - \sigma_3}{2} = \left(\frac{\sigma_1 + \sigma_3}{2} + \frac{\tau_0}{\tan \phi} \right) \sin \phi \qquad [6.5]$$

After rearranging the terms:

$$\sigma_3 = \sigma_1 \frac{(1 - \sin \phi)}{1 + \sin \phi} - \frac{2\tau_0 \cos \phi}{1 + \sin \phi} \qquad [6.6]$$

Since τ_0 and ϕ are constants, [6.6] could be written in the form:

$$\sigma_1 = a + b\sigma_3 \qquad [6.7]$$

where:

$$a = 2\tau_0 \sqrt{b} \qquad [6.8]$$

$$b = \frac{1 + \sin \phi}{1 - \sin \phi} \qquad [6.9]$$

Hence the results of the triaxial tests could either be represented by the

Mohr diagram, where τ_0 and ϕ are constants, or in the form of a linear graph between σ_1 and σ_3 in accordance with Coulomb's law (Fig. 108).

THE EFFECT OF FLUID-FILLED PORES ON THE STRESS DISTRIBUTION IN ROCK MASSES

Rocks within the outer few kilometers of the lithosphere have either an intergranular or a fracture porosity, and below the depth of a few tens of meters the pore spaces are filled with water or, exceptionally, with gas or oil. The pressure of water as a function of the depth Z is approximately given by the relation:

$$p = \rho_w gZ \tag{6.10}$$

where ρ_w is the density of water, and g the acceleration due to the gravity. However, as pointed out by Hubbert and Rubey (1959), the hydrostatic pressures encountered in the wells are markedly different from the pressures given by [6.10], and, at times, are found to approach the magnitude given by:

$$p = \rho_b gZ \tag{6.11}$$

where ρ_b is the bulk density of the water-saturated rock. The pressure given by [6.11] is equivalent to the entire weight of the overburden and is commonly known as the geostatic pressure. The interstitial water pressure exerts a buoyant force on the porous media in accordance with the principle of Archimedes. As discussed by Hubbert and Rubey (1959), the total force F exerted on the fluid-filled, porous media is equivalent to the force components F_s for solid and F_1 for liquid spaces:

$$F = F_s + F_1 = -\iiint\limits_V (\text{grad } p)\, dV$$
$$\tag{6.12}$$
$$= -\iint\limits_A p\, dA$$

In accordance with [6.12], this force may be computed by either integrating the fluid pressure p over the entire external area A of the space, or by integrating the gradient of the pressure with respect to the volume over the total volume V of the space considered.

Let us now investigate the effect of pore pressure on the stresses in a fluid-filled, porous medium. Consider a porous, fluid-filled specimen en-

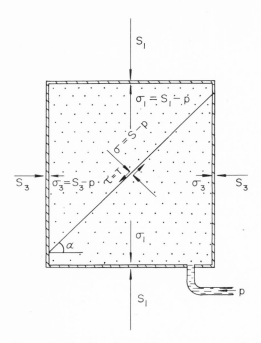

Fig. 109. Total and partial pressures on a jacketed specimen with internal fluid pressure.

closed in a flexible, impermeable jacket, subjected to a triaxial test. The total stresses S_1 and S_3 are applied externally, while a pressure p of arbitrary magnitude is added internally (Fig. 109). In accordance with [6.1] and [6.2], the normal and the shear components of the stress at any internal plane making an angle α with the S_3-axis would be:

$$S = \frac{S_1 + S_3}{2} + \frac{S_1 - S_3}{2} \cos 2\alpha$$

$$T = \frac{S_1 - S_3}{2} \sin 2\alpha$$

[6.13]

The external stresses, however, would be opposed by the pore pressure p which would completely permeate the solid and fluid-filled spaces within the jacket. Therefore, of the applied external stresses S_1 and S_3, only $(S_1 - p)$ and $(S_3 - p)$ would be effective in producing the deformation.

Consequently the normal and the shear stresses would be given by:

$$S' = \frac{(S_1 - p) + (S_3 - p)}{2} + \frac{(S_1 - p) - (S_3 - p)}{2} \cos 2\alpha$$

[6.14]

$$T' = \frac{(S_1 - p) - (S_3 - p)}{2} \sin 2\alpha \qquad\qquad [6.15]$$

On simplification we have:

$$S' = \frac{S_1 + S_3}{2} + \frac{S_1 - S_3}{2} \cos 2\alpha - p = S - p \qquad [6.16]$$

$$T' = \frac{(S_1 - S_3)}{2} \sin 2\alpha = T \qquad\qquad [6.17]$$

By definition, we have:

$$S_1 - p = \sigma_1, \qquad S_3 - p = \sigma_3, \qquad T = \tau$$

and hence it follows that, as a result of the reduction of the external stresses by the opposing pressure p, the normal and shear stresses inside the specimen become:

$$S' = S - p = \sigma \qquad\qquad [6.18]$$

$$T' = T = \tau \qquad\qquad [6.19]$$

This indicates that the deformation is caused by the stress having the components σ and τ, where the normal stress has been reduced while the shearing stress remains unchanged.

Some of the earlier tests confirming the theory developed in the previous sections were carried out by Handin (1958) on Berea sandstone. His results are shown in Fig. 110 where $\sigma_1 = S_1 - p$ values, corresponding to failure of

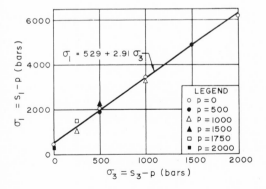

Fig. 110. Results of Handin's (1958) experiments on Berea sandstone with different values of internal pressure (pore pressure).

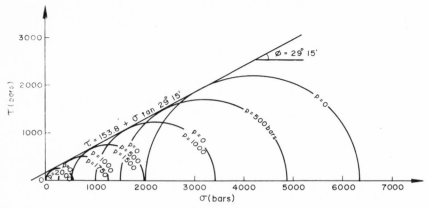

Fig. 111. Representation of Handin's (1958) test of Berea sandstone on the Mohr Diagram.

up to 6,300 bars, are plotted against $\sigma_3 = S_3 - p$ for pore pressures varying from 0 to 2,000 bars. Tests were performed for both the compressive and tensile stress S_1. The results are in good agreement with the values of $a = 529$ bars and $b = 2.91$. For these values of a and b, $\tau_0 = 154$ bars and $\phi = 29.25°$. Fig. 111 is a representation of these tests on the Mohr diagram. As confirmed theoretically and experimentally, the Mohr-Coulomb criteria for failure should be modified to:

$$\tau_{\text{crit}} = \tau_0 + (S - p) \tan \phi \qquad\qquad\qquad\qquad [6.20]$$

Hubbert and Rubey (1959) pointed out that, since Handin's tests have shown the validity of these relations to pore pressures of the order of 2,000 bars, which corresponds to the stress conditions existing at 20 km depth from the earth's surface, the conclusions drawn can be used confidently for the mechanical analysis of the tectonic deformation of porous rocks.

THE PART PLAYED BY FLUID PRESSURE IN OVERTHRUSTS

In their classical study, Hubbert and Rubey (1959) have shown the important part played by high pore pressures in understanding the mechanism of overthrusts where large masses of rock are displaced over a few kilometers at small dips (10° or less). Let S and T be the normal and shear components of the total stress across any given plane of a porous rock filled with fluid at a pressure p. Then the normal and shear components of the effective stress, in accordance with [6.18] and [6.19], are given by:

$$\sigma = S - p, \qquad \tau = T$$

Slippage along any internal plane in the rock would occur when the shear stress along that plane reaches the critical value in accordance with the Mohr-Coulomb law given by [6.20]:

$$\tau_{\text{crit}} = \tau_0 + \sigma \tan \phi \qquad\qquad\qquad [6.21]$$

After fracture has started, τ_0 becomes zero and further slippage would then occur when:

$$\tau_{\text{crit}} = \sigma \tan \phi = (S - p) \tan \phi \qquad\qquad [6.22]$$

If $p = \lambda S$, [6.22] could be written as:

$$\tau_{\text{crit}} = (1 - \lambda)S \tan \phi \qquad\qquad\qquad [6.23]$$

The two equations above show that without changing the coefficient of friction ($\tan \phi$), the critical value of the shearing stress could be made arbitrarily small simply by increasing λ, which increases with the pore pressure p. In the case of a horizontal block, the total weight per unit area S_{zz} is jointly supported by the fluid pressure p and the residual solid stress σ_{zz}. With the increase of p, σ_{zz} decreases, and when p approaches the limit S_{zz}, λ approaches unity and σ_{zz} reduces to zero. In effect, when λ approaches 1, the horizontal block is free to move laterally. In the case of gravitational sliding on a slope of angle θ, when T is the total shear stress and S is the normal stress on the inclined plane:

$$T = S \tan \phi \qquad\qquad\qquad [6.24]$$

substituting from [6.19] and [6.23]:

$$T = \tau_{\text{crit}} = (1 - \lambda)S \tan \phi \qquad\qquad [6.25]$$

Equating from [6.24] and [6.25] we have:

$$\tan \theta = (1 - \lambda) \tan \phi \qquad\qquad\qquad [6.26]$$

Eq. [6.26] shows that the slope θ, which the block could slide down, can be made to approach zero as λ approaches 1, which corresponds to the fluid pressure p approaching the total normal stress S.

Hubbert and Rubey (1959) cited a number of examples where pore-fluid pressures of the order of $0.9S_{zz}$ have been observed, supporting their hypothesis that, when the pore pressures are sufficiently high, very much longer fault blocks could be pushed over large horizontal distances or blocks under their own weight could slide down very much gentler slopes than would be otherwise possible.

THE EFFECT OF PORE-FLUID PRESSURE ON THE EFFECTIVE STRESS IN
DIFFERENT FAULT ENVIRONMENTS

In the previous sections we have seen that the increase of pore pressure
reduces the effective stresses and, in turn, can generate circumstances favor-
able for overthrusts and gravitational sliding on much gentler slopes than
otherwise required for such movements.

In a recent study, Snow (1972) showed how the creation of a reservoir
may produce the changes in the effective stresses. He carried out the analysis
for different situations of reservoir geometry and existing fault environments. For this analysis, he investigated the effect of plane-strain loading on
a model of orthogonal and deformable fractures and showed that, when
fractures are sound, tight and wide spaced (as must be the case in hypocentral regions at a few kilometers depth in unweathered surficial zones),
there could be large changes in the horizontal effective stress. We discuss his
analysis in some detail in the following paragraphs.

According to Snow (1972), in a region of recent faulting the most likely
existing stress state is near critical for renewed faulting. Under these circumstances the Mohr circle would be almost tangent to the failure envelope.
Hence, whether stability or instability accompanies the creation of a reservoir would depend upon whether the changes, brought about in the effective
stresses by the creation of the reservoir, drive the Mohr circle away or
towards the failure envelope. In accordance with [6.6], at failure:

$$\sigma_3 = \frac{\sigma_1 (1 - \sin \phi)}{1 + \sin \phi} - \frac{2\tau_0 \cos \phi}{1 + \sin \phi}$$

Differentiating σ_3 with respect to σ_1 we have:

$$\frac{d\sigma_3}{d\sigma_1} = \frac{1 - \sin \phi}{1 + \sin \phi} \qquad\qquad\qquad [6.27]$$

The condition, which is required to remain tangential to the failure envelope when the stresses change, is given by [6.27]. To find the result of man's
activity, the behavior of $d\sigma_3/d\sigma_1$ needs to be investigated for the different
boundary conditions that exist in the nature.

In the analysis carried out, it is assumed that there exists a principal stress
in the vertical direction equivalent to the overburden pressure. The effective
vertical stress would be equal to the overburden pressure minus the pore
pressure. Hence, if the pore pressure is known, the effective vertical pressure
could be found. Any change in the effective vertical stress would also change
the effective horizontal stresses. However, the changes in the effective horizontal stresses could be produced without changing the effective vertical
stress, for example, by the loading of an extensive reservoir, where a small

water head would change the effective vertical stresses negligibly compared to the changes in the effective horizontal stresses because of the changes in pore pressure due to large horizontal extent. The changes in the effective horizontal stress are governed by the deformability of the rock mass, including its faults, joints and inhomogeneities, as compressible elements. Generally a change in pore pressure does not affect the major and minor effective stresses by the same amount.

Theoretical and laboratory investigations on calculating the effect of pore-pressure changes on the effective stresses in fractured rock systems have shown that, depending upon the stiffness and orientation of the fractures, the lateral changes in effective stresses could correspond to the ones expected for values of Poisson's ratio, differing considerably from those assigned or possessed by the solid rocks investigated. According to Snow (1972), the existence of largely spaced, ($D \approx 10^2$ cm) stiff fractures ($C \approx 10^5$ bars/cm) in an area of an infinite reservoir created in a normal fault environment could favor failure when the product CD (in this case $CD = 10^7$ bars) is much larger than the modulus of the rocks, E, being of the order of 10^5 bars. However, he found that $CD = 3 \times 10^3$ bars from several field investigations (Snow, 1968a). For such small values of CD, the change in the effective stresses would be less and so would the chances of failure. However, detailed field and laboratory investigations need to be carried out for estimating the magnitudes of E and CD.

Let us now consider Snow's (1972) analysis of the change in the horizontal effective stress consequent to the change h in the water level of an infinite reservoir created over an axi-symmetric, permeable, fractured rock mass. Fig. 112 shows the vertical section of a fractured rock mass beneath an infinite reservoir, while Fig. 113 shows the deformation of a rock mass with deformable vertical fractures. Taking the compression as positive for stress and the extension as positive for strain, the strain component in the x-direction could be written as:

$$\epsilon_x = -\frac{1}{E}d\sigma_x + \frac{\nu}{E}d\sigma_y + \frac{\nu}{E}d\sigma_z \qquad [6.28]$$

where ν is Poisson's ratio and σ_x, σ_y, σ_z are the stress components in the x-, y-, and z-directions.

Under plane, vertical loading changes, no lateral deformation takes place in the mass containing deformable vertical fractures. Under these circumstances, as is clear from Fig. 113, the lateral strain will be:

$$\epsilon_h = \frac{dD}{D} = 0 \qquad [6.29]$$

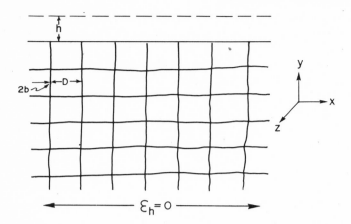

Fig. 112. Vertical section of a fractured rock mass beneath an infinite reservoir.

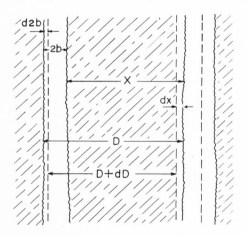

Fig. 113. Deformation of a rock mass with deformable vertical fractures.

and since $D = x + 2b$:

$$\epsilon_h = \frac{dx + d2b}{D} = 0 \tag{6.30}$$

where $2b$ is the width of the fractures spaced at D intervals. An increase of the aperture of the fracture and the dimension of the block would produce extension, which must be zero under plane strain, and hence:

$$dx = -d2b = \frac{1}{C} d\sigma_e \tag{6.31}$$

where σ_e is the contact stress; the fractures deform essentially linearly with increasing σ_e. The strain in the blocks alone is given by:

$$\epsilon_x = \frac{dx}{D} = \frac{1}{CD} d\sigma_e \qquad [6.32]$$

i.e. with the increase of the effective stresses the blocks extend while the fractures close. However, the total stress is responsible for the deformation of the blocks, and hence:

$$\epsilon_x = -\frac{1}{E} d\sigma_{tx} + \frac{\nu}{E} d\sigma_{ty} + \frac{\nu}{E} d\sigma_{tz} \qquad [6.33]$$

For symmetric fractures in the horizontal plane:

$$d\sigma_{tx} = d\sigma_{tz} \qquad [6.34]$$

and hence:

$$\epsilon_x = -\frac{1}{E}(1 - \nu)d\sigma_{tx} - \nu d\sigma_{ty} \qquad [6.35]$$

Since the change in the vertical load:

$$d\sigma_{ty} = \rho gh \qquad [6.36]$$

where ρ is the density and h is the change in height of the reservoir, we have:

$$d\sigma_{tx} = d\sigma_{ex} + \rho gh \qquad [6.37]$$

Upon filling an infinite reservoir, we have from [6.35], [6.36] and [6.37]:

$$\epsilon_x = -\frac{1}{E}[(1 - \nu)(d\sigma_{ex} + \rho gh) - \nu\rho gh] \qquad [6.38]$$

and substituting from [6.32], we obtain:

$$\epsilon_x = \frac{1}{CD} d\sigma_{ex} \qquad [6.39]$$

Hence:

$$d\sigma_{ex} = \frac{2\nu - 1}{(E/CD) + 1 - \nu} \rho gh \qquad [6.40]$$

Eq. [6.40] gives the change in the horizontal effective stress upon the change of the water-table level in an infinite reservoir (h), over an axi-symmetric, permeable, fractured rock mass after a long time.

Similarly it can be shown that, after a long period, the change in the horizontal effective stress upon the change in the water-table level (h) in an axi-symmetric, permeable, fractured semi-infinite rock mass is given by:

$$d\sigma_{ex} = \frac{\nu - 1}{(E/CD) + 1 - \nu}\rho gh \qquad\qquad [6.41]$$

The pore-pressure changes do not reach the basement rock instantaneously. During this transient period, before the arrival of the change in the pore pressure, there are effective-stress changes which are important. In the case of an infinite reservoir, as shown earlier, the strain of the solid block is given by [6.35]. Since $d\sigma_{ty} = \rho gh$ and $d\sigma_{tx} = d\sigma_{ex}$ before the arrival of the pore-pressure increase, we obtain on substitution in [6.35]:

$$\epsilon_x = -\frac{1}{E}[(1 - \nu)d\sigma_{ex} - \nu\rho gh]$$

$$= \frac{1}{CD}d\sigma_{ex}$$

and hence:

$$d\sigma_{ex} = \frac{\nu}{(E/CD) + 1 - \nu}\rho gh \qquad\qquad [6.42]$$

The change in effective stress during the transient period is given by [6.42]. Snow argued that the raising of the water table has no effect during the transient period since there is no increase in the total stress. He estimated the changes in the effective horizontal stresses, as a result of the filling of an infinite reservoir and the raising of the water table, for two sets of parameters: (1) for shallow, weathered rocks with closely spaced fractures, and (2) for deep, fresh, hard rocks with widely spaced fractures.

The results of his calculation, assuming that the porosity = 0, E = 3.5×10^5 bars and $\nu = 0.3$, are as follows: the change in the effective stress is represented by (Snow, 1972):

$$d\sigma_{ex} = \overline{K}\rho gh$$

where $\bar{K}$ = $\dfrac{v}{(E/CD) + 1 - v}$ during the transient period in an infinite reservoir

$= \dfrac{2v - 1}{(E/CD) + 1 - v}$ ultimate in an infinite reservoir

$= \dfrac{v - 1}{(E/CD) + 1 - v}$ due to the raising of the water table

in accordance with relations [6.42], [6.40] and [6.41].

1st case. For shallow, weathered rocks with closely spaced fractures, $C = 45$ bars/cm, $D = 25$ cm; hence $E/CD \approx 300$. Consequently, the values of $\bar{K}$ are those given in Table XVII. The calculations above show that under the assumptions made, the horizontal effective-stress changes are negligible (being of the order of 10^{-3} of the ρgh value) in weathered rocks.

2nd case. For deep, fresh, hard rocks with widely spaced fractures, the experimental value of $C \approx 10^4$ to 10^5 bars/cm, $D = 500$ cm; hence $E/CD \approx 2 \times 10^{-3}$ to 2×10^{-2}. Consequently, under these circumstances the changes in the effective stresses are considerable, as seen in Table XVII.

Snow (1972) also examined analytically the effect of filling an infinite reservoir as well as that of raising the water table in the environs of normal, thrust and wrench faults.

Normal faulting

In the regions of the normal faulting, the maximum effective stress σ_1 is vertical, and it remains unaffected by the filling of an infinite reservoir of height h; however, the horizontal effective stress reduces by $0.57 \, \rho gh$ (Table

TABLE XVII

Values of $\bar{K}$ in the relation $d\sigma_{ex} = \bar{K}\rho gh$ (after Snow, 1972)

Filling of an infinite reservoir		Raising of the water table
transient	ultimate	
1st case		
$\dfrac{0.3}{(E/CD) + 0.7} = 10^{-3}$	$\dfrac{-0.4}{(E/CD) + 0.7} = -1.3 \times 10^{-3}$	$\dfrac{-0.7}{(E/CD) + 0.7} = -2.3 \times 10^{-3}$
2nd case		
$= 0.43$	$= -0.57$	$= -1.0$

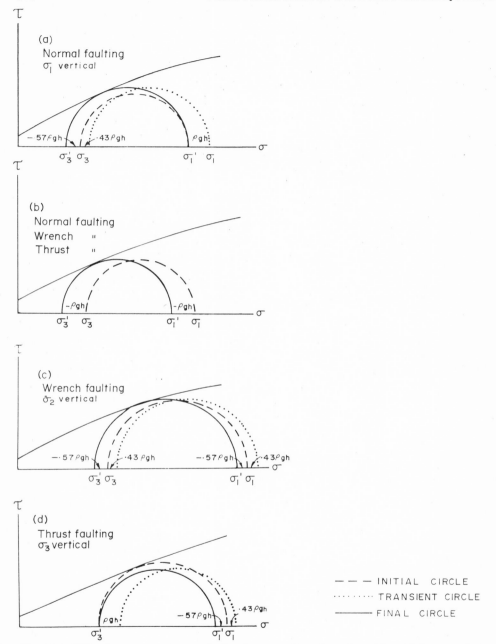

Fig. 114. Effective stress changes in hard, unweathered rocks with widely spaced fractures. (a) Filling of an infinite reservoir: normal faulting. (b) Raising of the water table: normal, wrench and thrust faulting. (c) Filling of an infinite reservoir: wrench faulting. (d) Filling of an infinite reservoir: thrust faulting.

XVII), thus driving the Mohr circle towards the failure envelope (Fig. 114a). In the case of a narrow reservoir, loading raises the water table causing a reduction of both the vertical effective stress and the horizontal effective stress by the same amount, i.e. ρgh (Table XVII), thus driving the Mohr circle towards failure (Fig. 114b). Such circumstances are often met within reservoirs filled for irrigational and hydroelectric purposes.

Wrench faulting

The stresses σ_1 and σ_3 are in the horizontal plane; hence, both are reduced by 0.57 ρgh when an infinite reservoir is filled, which drives the Mohr circle towards the failure envelope (Fig. 114c). Similarly, filling a reservoir on a narrow canyon would, in effect, raise the water table; thereby reducing the effective stresses σ_1 and σ_3 by ρgh, and thereby driving the Mohr circle towards failure (Fig. 114b).

Thrust faulting

In the environ of thrust faulting the maximum effective stress σ_1 is in a horizontal plane. Consequent to filling an infinite reservoir, the vertical stress σ_3 remains unchanged, whereas σ_1 is reduced by 0.57 ρgh, thus diminishing the diameter of the Mohr circle and consequently driving it away from the failure envelope (Fig. 114d). However, if an infinite reservoir created over a thrust fault is drained out, instability would be introduced. When a narrow reservoir is filled, and consequently, the water table is raised, instability would be introduced, reducing both σ_3 and σ_1 by ρgh (Fig. 114b).

EFFECTIVE-STRESS CHANGES DURING THE TIME TAKEN BY PORE-PRESSURE CHANGES TO REACH DEEP IN THE BASEMENT ROCKS

The pressure fluctuations caused at the surface as a result of changes in the reservoir water level affect the pore pressures at depth after a certain time lag. This is caused due to the storage capacity of the porous media. Howells (1973) estimated that, under certain conditions close to those observed in the field, the instantaneous variations of pressure at the surface would have a significant effect on the pore pressures at depths of 2.5—7.5 km after a lapse of some hundreds of days. As discussed earlier in the cases of reservoir-associated earthquakes, it has been indeed observed that the peak seismicity occurs after a certain time lag, following the peak reservoir levels in certain cases. Before the pore-pressure rise arrives, the total weight of the overlying reservoir loading is felt and an increase of effective stresses, both vertically and horizontally, occurs in accordance with Poisson's ratio. In a normal fault environment this gives rise to stability (Fig. 114a), but conversely, a lowering of the reservoir level could trigger failure. Considering the relationship:

$$\frac{d\sigma_3}{d\sigma_1} = \frac{1 - \sin \phi}{1 + \sin \phi}$$

which should hold good for the Mohr circle to move parallel to the failure envelope of the slope ϕ, the transient would produce (in accordance with Table XVII):

$$\frac{d\sigma_3}{d\sigma_1} = \frac{0.43 \, \rho g h}{\rho g h} = 0.43 = \frac{1 - \sin \phi}{1 + \sin \phi}$$

or, $\phi = 23.3°$. Since most fractures have a friction angle close to $30°$, failure would occur on decreasing the effective stress. In the case of the basement rocks stressed appropriately to cause wrench faulting, a positive propagating transient would increase both the horizontal effective stresses by $0.43 \, \rho g h$, and so stabilize the situation. The converse is true when the reservoir level is lowered. Similarly, a transient propagating in thrust environments would increase σ_3 by $\rho g h$ and σ_1 by $0.43 \, \rho g h$, thus rendering a stabilizing effect (Fig. 114d). Transients are not felt in deep regions which are unaffected by pore-pressure changes in all cases of a rising water table in low-porosity rocks, and hence there is no effect on the subsurface stress distribution.

PORE-PRESSURE CHANGES AND EARTHQUAKES

After having considered the effect of creating water reservoirs on the subsurface effective stresses under certain assumptions, we now discuss the possible part played by the pore-pressure changes in generating the earthquakes at Denver, Rangely, Kariba, Kremasta and Koyna.

Denver

Fractured crystalline rocks constitute the basement at a depth of 3670 m in the Denver Basin. Significant fluid loss experienced during coring and testing indicated that the natural formation pressure was much less than the hydrostatic head to the surface (by means of the drill-stem test it was measured to be 60 bars less).

Fluid injection started in March, 1962, when 1.6×10^4 m^3 of fluid was injected. The maximum wellhead injection pressure was 38 bars. No injection was made from September, 1963, to March, 1964. Injection under gravity discharge was then resumed and, on an average, 9.1×10^3 m^3 of fluid was injected monthly until March, 1965. Pumping pressures were then increased to about 70 bars in order to increase the discharge rate to about its initial value. The operation was suspended in February, 1966.

Healy et al. (1968) were able to show, through the analysis of the records of pumping pressure and injection rate, that the basement reservoir rocks were stressed to near-failure strength prior to the fluid injection. Snow (1972) has shown that the area was critically stressed due to the erosional unloading which had occurred since the Pliocene. Injection into the well had raised the bottom-hole fluid pressure. Correspondingly, the effective normal stress across potential shear fractures near the well bore was decreased within the rock, which was already stressed close to the point of failure, and hence earthquakes resulted. In the following we discuss their analysis.

Fault-plane solutions determined from P-wave first motions indicated a right-lateral strike-slip motion on vertical fault planes aligned with the trend of the epicentral zone. The presence of vertical fractures in the basement rock prior to injection, as found from cores and the alignment of one of the two possible fault planes according to mechanism solutions along the trend of the seismic zone, suggests that the earthquakes are produced by a regional stress field of tectonic origin.

According to the theory of faulting, strike-slip faulting occurs where the maximum principal stress, S_1, and the least-principal stress, S_3, are horizontal and the intermediate principal stress, S_2, is vertical. This intermediate principal stress is equal to the lithostatic pressure of 830 bars at 3,670 m below the surface at Denver. A pressure gradient of 0.226 bar/m for saturated sedimentary rocks was taken in arriving at this figure. The maximum principal stress, S_1, would therefore be at least 830 bars.

A linear pressure increase of about 7 bars per 379 liter/min increment was required to increase the injection rate at Denver to 1,414 liter/min. A linear extrapolation of the bottom-hole pressure from these data gives a pressure of 362 bars at zero injection rate (taking a pressure gradient of 0.0979 bars/m for the fresh water used in this test). This is the critical pressure, P_c, at which the fractures in the formation parted to accept fluid at large rates. From the theory of hydraulic fracturing (Hubbert and Willis, 1957), the hydraulic fractures in the well open normal to the direction of the least-principal stress S_3. In this particular case, S_3 being equal to P_c, it would have a value of 362 bars.

From the rate of the falloff of the fluid level in the well after fluid injection had ceased, a static head of 900 m below the well mouth has been estimated (Healy et al., 1968). This would make the initial fluid pressure (P_0) in the reservoir 269 bars before the injection of fluid at 3,671 m depth.

The pore pressure in the reservoir at the time when faulting was initiated (P_f) and the first earthquakes were recorded has been estimated to be 389 bars, as an excess pressure of 30 bars at the wellhead was required to achieve the same volume of fluid injection.

Given S_1 = 830 bars, S_3 = 362 bars, and P_0 = 269 bars, the effective shear and normal stresses on a potential fault plane have been found to be $\tau = 203$ bars, $\sigma = 210$ bars using formulae [6.1] and [6.2], [6.18] and [6.19]. The angle has been taken to be 60°, the value found typically for experimental shear fractures. Prior to injection, the ratio of shear stress to normal stress, τ/σ, was therefore 0.95, being slightly above the static friction value for most rocks (Byerlee, 1968). According to the Mohr-Coulomb criterion for failure, resistance to the fracture is given by [6.3]. From the above-determined values of the shear and normal stresses, and taking $\phi = 30°$, the value of τ_0, to prevent fracture works out to be at least 82 bars in the reservoir rock prior to injection. During fluid injection, when the pore pressure was raised to 389 bars, τ would be 203 bars and σ would be 90 bars. The occurrence of faulting upon reduction of the frictional term $\sigma \tan \phi$ by 69 bars indicates a value for τ of 151 bars or less. The cohesive strength τ_0 of 150 bars seems a reasonable assumption for fractured crystalline rocks. As the effective normal stress in the rock, which was already stressed so close to the point of failure, decreased by more than 50% as a result of fluid injection, earthquakes resulted.

This is consistent with Hubbert and Rubey's (1959) concept of earthquake triggering, but this experiment did not provide an adequate test of that hypothesis as the state of stress, or pore pressure, was not measured in the epicentral zone. In the Denver area, many of the earthquakes occurred 5 km away from the well. Moreover, three earthquakes of magnitude 5, the largest earthquakes in the sequence, occurred a year after fluid injection had ceased. The locations of these were also about 5 km west-northwest of the well. These observations were difficult to explain as it could not be estimated by how much the pore pressure had increased at a distance of 5 km after injection of 160 million gallons of fluid into the well. Presumably, a fracture zone along the epicentral trend, which alone constituted the available pore space in the basement rock, existed prior to injection. It was not possible to estimate the fluid storage capacity of the fractures as their extent and orientation was not known. Even if a detailed knowledge of reservoir rock characteristics had been known, they would have been of no use as the analytical tools are not yet available for a precise calculation of the storage capacity.

Rangely

The Rangely oilfield experiment provided the first adequate field test of Hubbert and Rubey's hypothesis where measurements of fluid distribution and rock stresses away from the injection wells have been carried out. The Rangely experiment was so designed that the fluid pressure could be reduced in a limited area of the reservoir rock where earthquakes were occurring. If the fluid pressure and the state of stress were known and the strength of the rock measured, Hubbert and Rubey's theory could be tested. The critical pore pressure below which no earthquakes took place could be compared with the pore pressure calculated theoretically to be critical for shear failure along a pre-existing fault.

   In the Rangely area, as a result of the water-flooding activity, the fluid pressures in the Weber sandstone along the periphery of the field have increased from the original normal hydrostatic pressure of 170 bars to 240—275 bars. Earthquakes between October, 1969, and November, 1970, occurred on the southwest extension of the fault where the fluid pressures were greater than 200 bars. Fault-plane solutions determined from the first motions of P waves indicated a strike-slip motion on the vertical faults trending either NW or NE (Raleigh et al., 1972). The NE-trending nodal planes are parallel to the mapped faults and to the trend of the epicenters, and hence are inferred to be the fault planes. The earthquake closest to the borehole in which the stresses were measured by hydraulic fracturing has a fault plane striking N50°E and dipping 80°NW. The maximum principal compressive stress in this solution is near horizontal and acts in an E—W direction. The slip direction plunges 20° towards N234°E. This inferred stress orientation is consistent with the stress measurements at Rangely. Surface measurements of the stress by De la Cruz and Raleigh (1972) at three sites gave the orientations of the maximum principal compressive stress as being N87°W, N83°W and N68°E. The magnitudes of the stresses, being a few bars, are small, indicating a significant relief of the stress by weathering, jointing, and other surface processes. Measurements of the state of stress in the reservoir rock (Weber sandstone, at a depth of 1,830 m) near the fault were conducted in 1971 (Haimson, 1972; Raleigh et al., 1972). An intact section of a new well was packed off and hydraulically fractured for this purpose. From the hydraulic fracture the values of 328 bars for the breakdown pressure and 314 bars for the instantaneous shut-in pressure (ISIP) were obtained. The orientation of the induced hydraulic fracture was measured to be vertical and striking N70°E. Using these values, and the value of 138 bars for the tensile strength of the rock obtained from the laboratory experiment, a complete determination of the magnitudes and orientations of the principal stresses was obtained (Raleigh et al., 1972). The total principal stresses thus calculated are:

S_1 = 590 bars N70°E horizontal

S_2 = 427 bars vertical (assuming that the lithostatic pressure is 0.23 bars/m depth)

S_3 = 314 bars N160°E horizontal

   Once the orientation of the fault plane and the slip direction in that plane is known, the shear and normal stresses, resolved along the slip direction and normal to the plane, can be calculated. The principal stress directions are

rotated to the new coordinate system with the following axes: (1) normal to
the fault; (2) in the slip direction of the fault; and (3) normal to the slip
direction, as inferred from the focal-mechanism solution. Thus, the normal
stress S obtained across the fault plane was 347 bars and the shear stress τ
along the slip direction was 77 bars.

According to the laboratory experiments by Byerlee (1971), slip should
take place in the Weber sandstone when:

$$\frac{\tau}{S-p} = 0.81$$

With the above values for τ and S, a pore pressure of 252 bars satisfies this
equation. This pressure is very near the bottom-hole pressure of 275 bars
during the injection when earthquakes were most frequent. After a month of
back-flowing of the four experimental wells, the bottom-hole pressure
dropped by 35 bars, and the earthquakes in the vicinity of the wells ceased.
The theory of Hubbert and Rubey (1959) is therefore in good agreement
with the observations.

Dieterich et al. (1972) have predicted the frequency and magnitude of
future shocks (if injection is resumed) by simulating the field conditions and
by applying the finite-element model of faulting to these simulated field
conditions.

Kariba

The three possible mechanisms for rock failure which could account for
the Kariba earthquakes following the reservoir impounding, are, as identified
by Gough and Gough (1970b): (1) the direct effect of the shear stress caused
by the load across the fault planes, (2) an indirect effect of the added stress
in triggering a larger initial stress, and (3) the effect of the increased fluid
pressure in the underground water.

A normal fault environment in the Kariba region is suggested by its geology and this is supported by focal-mechanism studies of the Kariba earthquakes (Sykes, 1967; Gupta et al., 1972b). According to Snow (1974),
under these circumstances there is little likelihood of failure by mechanisms
(1) or (2). Gough and Gough (1970b) also ruled out mechanism (1). If the
stresses are close to failure, what is required is a small change in the effective
stresses so that the Mohr circle moves towards the failure envelope (Fig.
114a, b). Failure could be caused by increasing the total vertical stress, if the
angle of internal friction of the fault is less than 23° for a value of Poisson's
ratio of 0.3. If Poisson's ratio is higher than 0.3 the friction angle would have
to be even smaller. As seen in Fig. 115, all the expectable combinations of
these parameters suggest that the field of instability is attained upon the

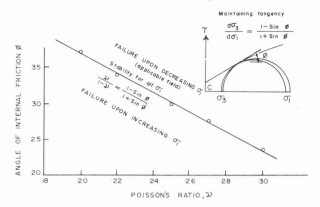

Fig. 115. Delineation of the combination of the frictional angle and Poisson's ratio producing failure upon increase, or decrease of effective stress.

decrease, not the increase of stress. This fact strongly supports mechanism (3).

The tremors in the Kariba region could be divided into three groups. The first group of tremors which occurred near Binga (Fig. 47a) during the filling of the lake in 1961—1962 is attributed by Gough and Gough (1970b) to mechanism (2) and/or (3). The Binga Fault was submerged by June, 1961, and some increase of fluid pressure must have occurred. Out of the reported 192 earthquakes in this group ($M_L \leqslant 3.2$), 180 occurred during a period of 4 days. The reservoir level had risen slightly above the river level at Binga when the earthquakes began. The important conclusion drawn from this observation is that stress changes of the order of a few tenths of a bar could cause the tremors; hence, stresses at the fault must have been near critical.

The second group of tremors which occurred during 1961 between Kariba and Chirundu (Fig. 47b) had earthquakes of magnitude $M_L \leqslant 3.6$. The stress changes due to the lake load were 0.3 bar for the closest of the hypocenters in this group, and 0.06 bar for the farthest. Gough and Gough (1970b) suggest that these tremors were due to mechanism (2). They ruled out mechanism (3) because of the remoteness of the hypocenters from the reservoir. However, Snow (1974) has drawn attention to the existing normal faults and argued that when a conduit is subjected to an increase of hydraulic potential of 160 m potential changes should occur over large areas, since the geomorphological and geohydrological considerations indicate interconnections and flows to many kilometers in depth and width. Hence, for this group of the earthquakes, mechanism (3) should not be ruled out on the basis of remoteness alone.

The third group of earthquakes includes larger shocks originating in the region overlain by the Sanyati Basin, which is the widest and deepest part of the lake (Fig. 47b). As discussed earlier, the existence of several NE-trending

normal faults and thermal springs supports the hydraulic continuity to greater depths. Depending upon the heterogeneity of fluid conductivity, the potential increase at the base of the sedimentary sequence traversed by normal
faults should be 0.7 to 0.9 times the increase of the reservoir level.

The loading history of the Kariba region, as traced by Snow (1974), seems
to show a critically stressed normal fault environment. Under these circumstances, filling decreases the horizontal effective stress σ_3 by about half the
overload ρgh, thus shifting the Mohr circle towards failure (Fig. 114a). In
view of the above, it may be interpreted that with the rise of the lake water
level, the effective stresses were reduced and this was the main agent triggering the earthquakes in an already critically stressed environment. The deformation of a rock mass whose pores are relatively undeformable is substanti-
ated by the observed subsidence.

Kremasta

The Kremasta region lies between the ENE-trending Anatoli and Pec-
Scutari transform faults (Dewey and Bird, 1970, p. 2642). A conjugate set of
wrench faults in the Kremasta region has been inferred by Galanopoulos
(1965). Snow (1972) assigned a wrench-fault environment to the Kremasta
region on the basis of these and some other evidences which were discussed
in Chapter 3.

As mentioned in Chapter 3, the Kremasta region has not been specifically
active historically (1700—1965), whereas an island 100 km southwest of the
region has been very active. The Acheloos River area had three earthquakes
of a magnitude exceeding 6 after reservoir impounding began in July, 1965.
Not one such earthquake occurred during the period 1821—1964 in the area.
An increase in seismicity following the reservoir impoundment has been
established beyond doubt in this region.

Papazachos et al. (1968) attributed the main shock which occurred on
February 5, 1966, to the triggering of failure in pre-stressed rock by the dead
weight of the reservoir (10^9 tons) acting vertically on a crustal block on only
one side of a vertical fault. However, such an occurrence has been ruled out
by Snow (1972), since consequent to filling a reservoir, during the transient
period before the pore pressure reaches the hypocentral depth, the dead
weight of the impounded water alone can not bring about the necessary
changes in the effective stresses to produce failure.

On the basis of geological and geohydrological considerations, Snow
(1972) has inferred raised hydraulic potentials at greater depths as a conse-
quence of reservoir loading. Snow has also traced the loading history of the
Kremasta region from the available geomorphological evidence and interpreted them in terms of the changes in the effective stresses. On the basis of the
above investigation, it is suggested that the Kremasta area, prior to the filling
of the lake, was in a condition of no special seismic hazard and the fluctua-

tion of stresses during the recent past had been of a smaller magnitude than the change imposed by impounding the reservoir in 1965. The raising of the reservoir water level to 274 m from a tail-water elevation of 140 m created a new low in the effective-stress history in the region of wrench and thrust faults. The earthquakes associated with an apparently normal displacement could also be triggered by the decrease of the effective stress after reservoir impounding, in accordance with Fig. 114. Snow noted that the onset of the strong foreshocks took place when the historic loading was first surpassed in 1966. The release of seismic stress or plastic flow has already raised the threshold, since no earthquakes occurred while the reservoir level was 25 m higher than during 1966 when the first swarm took place. The level must now rise to about 270 m (as in 1971) for earthquakes to occur, in comparison with the 245 m of 1966. As discussed earlier in Chapter 3, the peak seismic activity correlates well with the peak reservoir levels.

Koyna

As discussed earlier, the seismic activity had increased considerably in the Koyna region following the impounding of the Shivaji Sagar Lake in 1962. The earthquake focal mechanism determined for the main Koyna earthquake of December 10, 1967, by different investigators (as discussed in Gupta et al., 1972b) suggests a strike-slip fracture along a steeply dipping plane. Balakrishna and Gowd (1970) have inferred a wrench fault environment for the region on the basis of the geology and macroseismic effects of the earthquake. According to Krishnan (1966), epeirogenic movements have occurred from Pliocene to Recent times along the west coast of India where Koyna is situated. The western coast might have faulted during the Late Pliocene to Early Pleistocene. The Western Ghats, including the charnockite massifs of the Nilgiris, Coorg and Travancore were uplifted during that period and tilted, thus producing a steep western slope and a gentle eastern slope. Kathiawar has been uplifted in the Middle or Upper Pleistocene as evidenced by the presence of Miliolite limestone of Early Pleistocene age at an elevation of 305 m above sea level. Mild earthquakes have been recorded occasionally along the western coast, Cochin and other places. According to Snow (1972), in any region evidencing recent faulting the most likely state of stress is near critical for renewed faulting. On the basis of an analysis similar to that carried out for the Kariba region by Snow (1974), it could be inferred that the Koyna region was critically stressed and failure could occur following the changes in stress due to increased pore pressures.

Balakrishna and Gowd (1970), on the basis of certain theoretical considerations, have shown that wrench shear fractures cannot be developed due to the effect of loading alone. They then showed how an increase in pore pressure can change the ratio of maximum to minimum principal stress to cause failure. For certain sets of λ (the ratio of pore pressure to normal stress), horizontal stress and Poisson-ratio values, they estimated the ratio of σ_1/σ_3 with depth and investigated the limiting depths in which wrench shear fractures would occur following the increase of pore pressure. Finally, having estimated the rock volume for the December 10, 1967 earthquake, they calculated the amount of water that must have percolated through the cracks and fissures of trap rock and reached the Dharwar schists during a period of a few years from the day of impounding until the occurrence of the moderate-magnitude earthquake in December, 1967. They felt that this calculated amount of water could have easily found access, keeping in view the fact that a comparable amount of water penetrated under its own gravity into the schists rock at the Arsenal disposal well at Denver in a period of about 32 months. From the above it seems that the fluid pressures in the Koyna region must have played an important role in causing the earthquakes following reservoir impounding. On the Mohr stress diagram the wrench failure in the Koyna region following the increase of pore pressure could be represented by Fig. 114c.

SOME RECENT DEVELOPMENTS IN PORE-PRESSURE STUDIES

Servo-controlled testing

Recent developments in experimental techniques have made it possible to investigate the magnitude and nature of the variation of pore volumes during the disintegration process of the rock specimens tested under triaxial loading. For such investigations it is necessary to avoid the sudden collapse of the specimen which normally occurs in the conventional testing machines when the axial load-bearing capacity of the specimen is exceeded. Cornet and Fairhurst (1972) have used two servo-controlled loops for such experiments. The predetermined constant value of the confining pressure is maintained throughout the experiment by arranging the first loop in such a way that any increase in the confining pressure causes a reduction of the applied axial load to the extent necessary to return the confining pressure to the present value. Similarly, any drop in the confining pressure causes an increase in the applied axial load. With the help of the second loop, the confining pressure fluid is withdrawn at a constant rate without changing the confining pressure. The removal of the confining pressure fluid by the second loop causes the first loop to change the axial load to laterally deform the specimen just enough to compensate for the fluid extracted. When the lateral deformation is too rapid the axial load decreases. Thus it has been possible to deform the specimens to their complete disintegration in a controlled way.

Cornet and Fairhurst (1972) applied the above-mentioned experimental technique to test the validity of the "effective-stress" concept and to investigate the variations of pore volume associated with different modes of failure.

The experiments were conducted on saturated specimens of Berea sandstone. The effective-stress concept is found to be applicable to the phenomenon of disintegration, since it was found that the stress concentrations are functions of the effective-stress tensor. At low values of confining pressure, when the pore pressure was held constant, disintegration was found to be characterized by a continuous pore-volume increase and it was associated predominantly with vertical splitting and slabbing. At intermediate confining pressure and constant pore pressure, the pore volume is found to increase initially and then decrease, from which it could be inferred that the disintegration was initiated by the vertical splitting changing later to a predominantly shear phenomenon. At higher confining pressures, the shear phenomenon dominates, showing an overall decrease of the pore volume. Brace and Martin (1968) have shown that for rocks of low permeability, the strain rate at which the tests are performed have a significant influence on the peak strength. In light of the experiments carried out by them, Cornet and Fairhurst (1972) inferred that this strain rate effect is influenced by the confining pressures, since this parameter controls the mode of failure by governing the pore-volume changes.

Somewhat similar results have been reported by Goodman (1973) from their experiments on both intact and jointed specimens of friable Permian sandstone. The triaxial compression tests carried out on an intact rock specimen, which was initially saturated and undrained during the shearing, showed first an increase and then a decrease in the pore pressure. This change in pore pressure reflects the change in pore volume as suggested by Bieniawski (1967), Bruhn (1972), and others. At low confining pressures, the pore pressure at the time of rupture was lower than at the instant of initial deviatoric loading. Next, the jointed specimens were tested under triaxial loading. These tests indicated a relatively higher pore-pressure increase on loading, and the slip occurred when the pore pressure was at its highest level, suggesting that the pore pressure measured in a fault may be elevated before slip. Rummel and Gowd (1973) have reported results of triaxial tests on Ruhr sandstone specimens having intermediate porosity (2— 3%) and permeability (about 5 millidarcy). The confining pressures were kept constant during the test and were varied up to 1 kbar. The maximum load-bearing capacity of the rock was found to be a linear function of the confining pressure. At low strain rates, the law of effective stress holds good, whereas an apparent increase in strength is observed for higher axial strain rates. This discrepancy has been explained as due to the effect of dilatation during the progressive fracture. The dilatancy phenomenon as described by Scholz et al. (1973) is as follows. A water-saturated rock under increasing strain reaches a stage where the opening of new cracks cannot be matched by the flow of fluid into these cracks. The pore pressure thus drops and so does the compressional wave velocity. This process, however, leads eventually to the rock being hardened and at this stage no more cracks form, but the water continues to flow into the unsaturated region. This increasing pore pressure (with the compressional velocity returning to normal) leads finally to an earthquake.

Fairhurst (1973b), in his latest study, mentioned that the "effective-stress concept", which is applicable in the early stages of deformation, does not amply describe the transient pore-pressure conditions, which develop during the microcracking and subsequent stages of deformation. Macroscopic shear-plane development causes a reduction of stress in the unfailed rock adjacent to the planes and the associated local increase of pore pressure. According to Fairhurst, the presence of fluids under pressure thus enhances unstable deformation and failure processes.

In-situ measurements of stresses

The in-situ stresses at considerable depth can be measured by the technique of hydraulic fracturing of a borehole (Hubbert and Willis, 1957; Haimson and Fairhurst, 1970; Fairhurst and Roegiers, 1972). The maximum and least-principal stresses can be measured if the rock is not already fractured, and its tensile strength is known. The technique of hydraulic fracturing consists of sealing off an interval in the borehole with inflatable packers, in which the fluid pressure is raised by pumping from the surface until a sudden pressure drop occurs corresponding to the opening of a tensile fracture in the wall rock. The bottom-hole pressure, P_f, at which the breakdown occurs, is measured. Fluid is then pumped in rapidly to open and extend the fracture. The hydraulic fractures may extend to 100 m with a width of about 1 cm. After fracturing, sand, or any other propping agent, is pumped with the pressurized fluid into the fracture to hold it open after the pressure is released. After the pumps are shut off the pressure drops suddenly to a value called the instantaneous shut-in pressure (ISIP) which is equal to the least-principal stress, S_3. This pressure remains nearly constant until the surface pressure is released.

The orientation of the fracture is determined by an impression packer which has an impressionable rubber wrapped around a mandrel. The packer is lowered to fracture depth and the rubber membrane is forced against the borehole under high fluid pressure. The pressure is released after a few minutes, then the packer is deflated and removed from the hole. The impression of the fracture is clearly recorded on the rubber membrane and gives the orientation of the fracture.

Theoretically (Hubbert and Willis, 1957; Haimson and Fairhurst, 1970) the bottom-hole pressure, P_f, at which hydraulic fracture occurs is given by:

$$P_f = T + 3S_3 - S_1 - P_0$$

where P_0 is the interstitial fluid pressure of the rock measured before the pumping of the fluid, T is the tensile strength, and S_1 and S_3 are the greatest and least-principal stresses (compression is taken as positive). The tensile fracture extends in a plane normal to the direction of the least principal stress existing in the rock mass. If the vertical, or lithostatic pressure, is less than the horizontal compression stresses then the fracture will extend in a horizontal plane, and if the lithostatic pressure exceeds the horizontal stresses then the fracture will be vertical. The least-principal stress is given by the value of the ISIP. The tensile strength of the rock is measured in the laboratory. One other condition which needs to be satisfied for calculating S_1 from the above equation is that the fluid should not have permeated into the wall rock.

All the provisions for these measurements have been put in a single assembly, called the "deep stress probe" (Fairhurst and Roegiers, 1972). This consists of: (1) packers to take the impression of the wall before and after the hydraulic fracturing, (2) straddle packers to seal off the interval to be fractured, and (3) bottom-hole clockwork pressure recorders, to record the fluid pressures at all stages of the operation.

Coupled stress—flow method of analysis

In the recent years, having realized the important part that fluid-pressure variations play on the subsurface stress distribution, considerable effort is being directed to simulate models which would eventually approximate the subsurface conditions and study the pressure distribution following the fluid injection and its effect on the ambient stresses.

The basic problem in estimating pore pressures in the fractured reservoir rocks has been the sensitivity of permeability to the crack width, and hence to the effective stress across the fracture. In recent years, efforts have been made towards solving this problem. Rodatz and Wittke (1972), Noorishad et al. (1972), and Morgenstern and Guther (1972) have given sophisticated theoretical methods of calculating pore pressure and stress changes in fractured rock systems. The special significance of their work is that they have coupled the interdependent influences of pore pressure and rock stress using the technique of finite-element analysis.

In order to determine the steady-state distribution of pore pressures when fluid moves through a deformable mass of fractured rock, the methods of Morgenstern and Guther (1972) and Noorishad et al. (1972) employ a finite-element program for fluid-flow analysis coupled to a finite-element program for stress analysis through an empirical relation between the permeability and effective stress. A geometry of the fractured system, i.e. orientation, spacing and distribution of apertures, and the mechanical properties of both the intact rock and the rock fractures are assumed for these calculations. In the sample calculations, some rather simple systems have been analysed so far.

In these numerical calculations, the essential assumption is that the relationships between the crack width and effective stress and between the effective stress and permeability are either known or measurable. The experimental results of Jouanna (1972) have shown that these measurements are possible. The permeabilities observed in the field are found to be greater by several orders of magnitude than the laboratory value. This discrepancy is due to the presence of cracks and discontinuities in the rock mass which offer much less resistance to flow compared to the resistance offered by the microcracks and pores in the intact rock used in the laboratory tests. Various laboratory and field data have suggested that permeability reduces with pressure. From experimental evidence it has been suggested that the changes in effective stress are more dominant compared to the changes in shear stress. Hence, it has been assumed that permeability is a function of effective stress. With the help of existing laboratory and field data, Morgenstern and Guther (1972) have inferred that permeability increases semi-logarithmically with the decrease of effective stresses. Le Tirant and Baron (1972) suggested an exponential increase.

By assigning the geometry and permeability of individual elements in the fracture system it is possible to calculate numerically the pressure distribution, assuming a linear relationship exists between the flow rate and the pressure difference, i.e. a Darcy type of flow. For complex geometries and spatial variations of permeabilities, the finite-element method is the most useful (Zienkiewiez and Cheung, 1967). Rodatz and Wittke (1972), Noorishad et al. (1972) and Morgenstern and Guther (1972) have used the finite-element method, which depends upon dividing the fracture system into a network of fracture elements. Having obtained the distribution of the pressures at every point within the fractures, the equivalent point forces of these fluid pressures are calculated at the intersections, called nodal points, of the fracture elements. These forces are needed in the finite-element stress analysis for the whole rock mass.

Rodatz and Wittke (1972) have applied the fluid-flow analysis to both two- and three-dimensional problems. By coupling the interdependent effects of hydraulic and mechanical properties, the modifications of the permeability due to changing stresses by seepage flow and the decrease of strengths due to the interstitial water have been shown (Wittke, 1973). The coupled stress—flow analysis has been restricted so far to only the two-dimensional steady-state case (Noorishad et al., 1972).

It is assumed that the structure consists of a finite number of elements which are connected at a finite number of nodal points. In order to facilitate the exchange of information between the two computer programs for flow and stress analysis, identical grids are used for both. The grid for the stress analysis is taken in the same way as that for the fluid-flow analysis, except that the fractures here are no longer fractures but only a means of dividing the structure into finite elements.

With the nodal-point forces obtained from the fluid-flow analysis the effective stresses are calculated using the known formulae, as given by Noorishad et al. (1972) and Morgenstern and Guther (1972). The effective stresses are then used to find the permeabilities.

Since the effective stresses in the medium are dependent on the water-pressure distribution, and the pressure distribution is itself dependent on the stresses involved, an iterative technique, coupling both finite-element programs, is needed. A trial flow analysis is undertaken and the first approximations of nodal-point forces are obtained for use in the stress analysis. The effective stresses are then computed and used to find the permeabilities. The permeabilities are then assigned to the flow elements and the pressure distribution is found by iteration. The coupled programs iterate until no significant changes in the pressure distribution are found.

The mechanical effects of fluid in fissures are significant for problems concerning the failure of slopes and dam abutments, and for determining the permeability if hydraulic fracturing is utilized to increase the permeability for subsurface waste disposal and production of oil from reservoirs of extremely low permeability. The storage capacity of the reservoir rock can be assessed from these determinations. It has been seen that the storage and the transmissibility of the fractured reservoir rock would change as a result of the seepage flow. In a sample calculation of Noorishad et al. (1972), the injected fluid flows in two orthogonal cracks, which intersect at the well. The opening of these two fractures imposes an increased normal stress across all the other fractures in the system, causing a decrease in the permeability and fracture porosity throughout the reservoir. However, the opposite effect would be seen if a hydraulic fracture was opened along a plane perpendicular to the least-principal stress. In this case, as the fluid pressure rises in the hydraulic fracture, the orthogonal intersecting fractures will experience a decrease in the effective normal stress, and hence an increased permeability results.

This method of coupled pore-pressure—rock-stress analysis seems promising, and it looks possible to determine the changes in water pressures and stresses in a reservoir rock as a result of the seepage flow. The other factors, besides the fluid pressure, which may possibly affect the stresses are: (1) the heat transfer and the resulting thermal stresses, and (2) the effect of pressure gradients on the total stress field. Research is under progress to study these effects and to translate them in numerical code.

Chapter 7

FURTHER TOPICS

DAM SITE INVESTIGATIONS

Detailed geological maps are usually prepared around the foundation of a proposed dam. In the case of a large reservoir (volume $\geqslant 10^9$ m^3, usually impounded behind a dam of height $\geqslant 100$ m), it is desirable to carry out detailed geological mapping for the entire reservoir area (National Academy of Sciences, U.S.A., 1972). The competence of the rocks in the reservoir area and the investigation of the nearby faults and their hydrological significance deserve special attention. In the case that large faults evidencing recent movements are delineated, it is desirable to re-examine the suitability of the chosen site and look for possible alternative sites.

During its first meeting held at Paris on 14—16 December, 1970, the UNESCO Working Group on "Seismic Phenomena Associated with Large Reservoirs" recommended that the instrumental studies and surveys at the sites of large reservoirs be planned in the following two phases:

Phase 1, coinciding with the preliminary or feasibility stage of the project, should include:

(1) A study of the historical seismicity of the reservoir region, i.e. within its tectonic setting.

(2) A preliminary geological and geomorphological survey of the reservoir and its adjacent area, aimed at identifying potentially active geological structures.

(3) Sampling of seismic conditions. This can be carried out using a tripartite radio-linked high-frequency portable network, the purpose of which would be to monitor the pre-impounding seismicity as well as to establish the areas of minimum noise level prior to the start of construction.

If, on review of the information provided by Phase 1, it is concluded that such studies of the reservoir should be intensified, they may be continued in *Phase 2*, which should commence one or two years before impounding and should include:

(1) A more detailed geological and neotectonic survey of the reservoir and its adjacent area.

(2) The installation of permanent seismological instrumentation.

(3) Possible other work such as: (a) a program of precise levelling, (b) instrumentation to detect the activity of faults, and (c) studies of the stability of the reservoir slope.

Some details for seismic surveillance and geodetic measurements at the dam sites are given in the later part of this chapter.

In addition to the above-mentioned dam-site investigations, it is worthwhile estimating the in-situ stresses by means of the hydraulic fracturing technique wherever possible. Relevant details of the technique are given in Chapter 6. Fairhurst (1973b) has mentioned that "deep stress probes" have been successfully used to measure stresses at a depth of 305 m in granite and it would be soon possible to measure stresses at depths of 1524 m and even deeper. Obviously, the sites showing underground rocks which are near-critically stressed should be avoided for dam construction.

SEISMIC INSTRUMENTATION OF DAM SITES

With the exception of a few cases where seismic stations existed by chance, or where the Project Managers had the foresight to install the seismographs before impounding large artificial reservoirs, well-documented records of the change caused in the regional seismic status by creating artificial lakes are rather rare. In most of the cases it is found that the seismic stations were commissioned when reports of felt earthquakes began to be prevelant following the filling of artificial lakes. It has been also noticed that in the absence of a sufficient number of stations suitably distributed with respect to the reservoir geometry and/or due to the lack of suitable seismographs, accurate time-keeping and other factors, such as knowledge of the velocity structure, it has not been possible to determine the focal parameters accurately enough to delineate the trend of the hypocenters and assess their association with the known regional geological features. It is interesting to cite the example of the Denver earthquakes in this connection. During the period when one to five seismic stations were used, the hypocenters were found to lie in an area extending 75 km in length and 40 km in width with focal depths of up to 40 km. Healy et al. (1968) reported that later, with the help of eight L-shaped arrays, each having six seismometers at ½-km intervals in a small area around the well, and with good velocity control, the hypocenters were located in an area of 8 km length and 2 km width passing through the disposal well with focal depths ranging between 4.5 and 5.5 km.

The importance of instrumenting the large dams has been now well realized, having understood that damaging earthquakes could occur following the impounding. The UNESCO Working Group on "Seismic Phenomena Associated with Large Reservoirs" has specifically stressed upon the installation of permanent seismological stations well in advance of reservoir impounding. It is desirable that these stations continue to operate for a few years after impounding even when a particular reservoir does not show any seismic effect. During the third meeting of the Working Group, held at London in March, 1973, Professor Rothé pointed out that, probably as a result of the earlier recommendations of the Working Group, many large reservoirs have been instrumented during the preceding two years and that there have been many inquiries from the Project Engineers regarding the type and the number of seismic stations that should be installed at the dam sites. Consequently the Working Group invited Drs. R.D. Adams, D.I. Gough and K.J. Muirhead to make detailed recommendations for suitably instrumenting the artificial reservoirs. In the following we make suggestions based on the recommendations of Adams et al. (1973) and our experiences of establishing and operating seismic stations.

Sources of error in calculating the focal parameters

Precise time-keeping is a prime requisite since it is used as the base for reading the onset times of the waves belonging to the P and S groups when calculating the epicentral distance and origin time, and for corroborating similar observations made at different seismic stations. For locating the hypocenters within a kilometer, it is necessary that the times be known to 0.1 second or better. Time accuracy of this order is easily achieved by using quartz crystal chronometers, which have a stability of 1 ppm and are readily available on the market these days, and by making daily calibrations against standard radio time signals. Preferably, the radio time signal should be directly impinged on the seismograms along with the clock signals.

Sometimes it becomes difficult to read the initial P onset time when the phase emerges slowly from the background noise instead of being impulsive. This is more true for the emergent S phases since their onset may be masked by the P-wave coda, especially the large events for which the P-wave coda saturate the recording system. To enhance the signal to noise ratio, it is desirable to choose a seismically quiet site away from natural and man-made sources of seismic noise. To facilitate the detection of the smaller local events, it helps to peak the instrumental response in the frequency range from 4 to 20 Hertz, depending upon the local geological characteristics. The two steps mentioned above are helpful in enhancing the signal to noise ratio, particularly for P waves. The enhancement of S-wave signals is relatively much more difficult. It helps to install two orthogonally oriented horizontal seismographs in addition to the vertical component for a better detection of the S wave. However, it is preferable to have a larger number of stations with only a vertical component rather than fewer stations with three components.

The absence of knowledge of the crustal layer velocities renders precise location of hypocenters impossible. This is especially true when the epicentral distances are less than 1,000 km, when the seismic rays travel through the crust over large portions of their total path, and the crustal velocities are found to vary significantly from one region to another. The crustal velocities and structure are best determined by conducting deep seismic soundings along three or four profiles within an area of a few tens of kilometers from the reservoir. However, this being an expensive proposition, an effective alternative is to install the seismic stations well in advance and analyze the records of explosions used for the construction of the dam. The data thus obtained could suitably be supplemented by a few timed or radio signal-triggered explosions at the reservoir site for obtaining the velocities in the crustal layers.

Number and location of seismic stations

We have seen in Chapter 3 that most of the earthquakes induced by the reservoirs lie in their close proximity, more than three-fourths falling within 25 km from the center of the reservoir. A majority of these are located close to the deepest portion of the reservoir. Hence, in the absence of any previous knowledge of the probable post-impounding activity, seismic stations should be suitably located so that they could best monitor the seismic activity close to the region of the computed maximum depression of the crust as a result of the reservoir load.

Installation of only one station gives an idea of the hypocentral distance using the S minus P travel times within which the foci must lie. Being closer to the foci helps in restricting the space to which the earthquake could be assigned. However, single stations, equipped with matched three-component seismographs, give an idea of the direction from which the waves arrive; hence, a very approximate location of the epicenter could be obtained using the Galitzin technique. Although the installation of two stations does not permit the unique determination of the epicenter, it helps in enabling real events to be distinguished from the noise. If only two stations are to be installed, they should be sited on the two sides of the reservoir. With three seismic stations, it is possible to locate the epicenter. These stations should be installed so as to form a triangular network having dimensions comparable to those of the reservoir. Such a network has good locating capabilities for the area falling within the network; however, outside the network the location capabilities deteriorate considerably.

A network consisting of three seismological observatories will monitor general earthquake activity. However, in order to obtain a reliable estimate of the focal depth and epicentral coordinates, at least five seismic stations are required. Consideration of the geometry of the location of the seismic stations with respect to the region whose seismicity is to be monitored is important in minimizing the error of the focal parameter. Sato and Skoko (1965) considered 1,000 possible distributions of five seismic stations confined to a small region and determined the errors of depth, epicenter location, origin time and velocity determination under certain assumptions to simplify the computations. These computations have been carried out under the assumption that the hypocenter was located in the center of the area. In

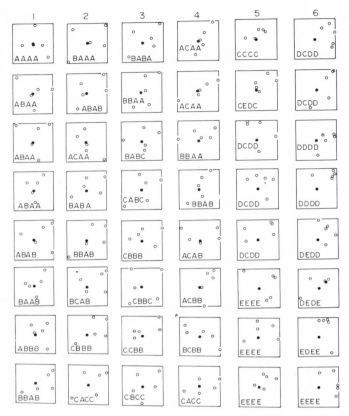

Fig. 116. The errors of depth, epicenter location, origin time and velocity are indicated by the capital letters at the bottom of the squares, where *A, B, C, D* and *E* represent very good, good, average, bad or very bad distribution, respectively (redrawn from Sato and Skoko, 1965).

Fig. 116 we reproduce some of their results, which fall under the following six broad categories.

(1) When three observation points form a triangle, and the focus and the other two observation points are inside, the result is generally good. Especially when three stations form a triangle, the fourth station is on one side and the fifth is close to the epicenter.

(2) When four observation points form a flat rhombus with the epicenter and the fifth point inside, the result is also good.

(3) When four observation points form a flat parallelogram and the focus and the remaining points are inside, the result is also good.

(4) When the observation points form a lens-shaped polygon with the focus inside, determination of the velocity is good.

(5) When the focus is on a side of a quadrilateral which is not flat, the result is not good.

(6) When the points are distributed on one direction of the focus, the result is generally not good.

The above-mentioned analysis is helpful in deciding the location of seismic stations with respect to the reservoir.

Care should also be taken to operate one or two sets of low-gain instruments which are useful in determining the magnitude of moderate or large events which may saturate the high-gain instrument records. It is also desirable to install at least two sets of three-component strong-motion accelerographs capable of recording accelerations in the 0.01—1.0-g range. These should preferably be installed in the body of the dam, somewhere close to its top and bottom.

It is desirable for the local seismic network, set up to monitor the seismic activity of a certain reservoir, to have liasion and exchange of data with the national network of seismographs if one exists in the country, or with nearby seismic stations equipped with sensitive short-period seismographs. Standby arrangements for the continuous operation of the seismographs by direct-current accumulators in an event of power failure is a must. In addition to the permanent stations, in the case a reservoir exhibits seismicity of appreciable magnitude, it may become necessary to install additional mobile stations to determine the focal mechanism solution and the relation of the hypocenters with the known or later discovered geological features.

It is necessary to keep a record of nearby explosions to prevent that these are mistaken for microtremors. Knowledge of the location and blasting time of the explosives could be very useful in checking the adopted location procedures and velocity models.

GEODETIC MEASUREMENTS

As discussed in Chapter 3, subsidence has been observed in the Lake Mead area following the lake's impounding (Rogers and Gallanthine, 1974). Gough and Gough (1970a) showed that the measured deflection along a re-levelled road agreed closely with the calculated elastic deflection. Adams et al. (1973) observed similar elastic deflection occurrences at the Libby and Dworshak reservoirs in the northwestern United States. Precise estimates of movements along certain well-known faults, such as the San Andreas Fault of California (Nason and Tocher, 1969), are now available. It has been also shown that the variation in movements at a particular place of the San Andreas Fault system preceded earthquakes of magnitude 4.5—5.5 (Hoffman, 1969). Geodetic measurements of crustal movement form an important part of the earthquake prediction programs now underway in the United States, Japan, and the U.S.S.R.

Geodetic studies before and after the filling of a reservoir, with special emphasis on the measurement of vertical displacements, could resolve, in case earthquakes happen to occur, whether they are caused by the load of the impounded water. Long geodetic level lines established before impounding and their repetition after impounding provides crucial information about the elastic deformation at the dam sites. Operation of continuously recording strainmeters makes an important addition to geodetic measurements. Tiltmeters for measurement in boreholes are now available. In the report of the National Academy of Sciences, U.S.A. (1972) the installation of tiltmeters in at least three widely spaced boreholes prior to the filling of the reservoir is recommended. However, it needs to be mentioned that the continuously recording strainmeter-type measurement is no substitute for the long level lines.

For constructional purposes detailed levelling is usually undertaken at almost all the reservoir sites. Adams et al. (1973) recommend that, in order to investigate the response of the crust to reservoir loading, levelling should be undertaken along at least one route beginning at the dam site and extending away from the future reservoir for about a length equal to one-half of the future reservoir length, or until it joins an existing levelled route extending to such a distance. They also recommend that at least one route should be re-levelled after the impounding to a distance of 50 km or more from any point on the reservoir periphery.

In the event of reservoir-inducing tremors after impounding, it is necessary to repeat the first-order triangulation surveys over and in the vicinity of the known geological faults in the project area. This should invariably include the epicentral region delineated by the seismic stations in operation. In addition to triangulation, precise levelling should be undertaken across several sections of the known geological faults and the trend of the epicenters delineated by the seismic stations.

ASEISMIC LAKES

To comprehend enhanced seismic activity following impounding, it is important to examine why a majority of large reservoirs do not show any seismic effect. During the first meeting of the UNESCO Working Group on "Seismic Phenomena Associated with Large Reservoirs" held at Paris in December, 1970, reference was made to the available information concerning about thirty large reservoirs. Impounding at only half of these lakes was accompanied and followed by seismic activity, i.e. by earthquakes whose foci appeared to be located in the close vicinity of the reservoirs and whose magnitude and frequency was higher than normal for the region.

Rothé (1968) has pointed out that, following the impoundment of reservoirs, the tremors occur under specific geological conditions only and the

building of dams does not always produce earth tremors. He has cited the
example of the Serre-Poncon Dam in the French Alps, where the artificial
lake is entirely situated on a flexible terrain constituted by black soil, in
which stresses were unlikely to be accumulated. No earthquakes have occurred in the region following reservoir impounding. Similarly Adams and
Ahmed (1969) attributed the less-pronounced seismic effect of the Mangla
reservoir to the 3 km thick layer of friable sandstones and clays of the
Siwalik (Late Tertiary) Series which would offer little resistance to any
applied load and which would deform gradually to accommodate it. They
observed that, in contrast to the situation at Mangla, at Koyna the lake is
formed on more competent basalt flows which could have a great amount of
latent seismic energy stored in them and this energy is ready to be triggered
by reservoir impounding. The aseismic behavior of the Bhakra Nangal Dam
in the foothills of the Himalayas in India, which has a much larger capacity
than the Koyna reservoir, could be comprehended in the same way since it is
also located over incompetent rocks.

This factor could be held responsible for leaving the seismic status of a
number of sites unaffected following the impounding of reservoirs. The discussions in Chapter 4 also show that initially competent rocks are more
suitable for the triggering of moderate-magnitude earthquakes.

Handin and Nelson (1973) have examined the aseismic behavior of Lake
Powell, which is located in Utah and Arizona in the southwestern United
States. Lake Powell was created in early 1963 by constructing the Glen
Canyon Dam on the Colorado River. It is a large, deep reservoir and by
September, 1972, it had accumulated about $2,000 \times 10^6$ m^3 of water with a
depth of 140 m near the dam. The lake is about 300 km long with an
extensive 3,000 km long shore line. Water in the lake is collected mostly over
Jurassic Navajo sandstone. Pre-impounding estimates of storage water loss,
based upon field and laboratory permeability measurements, are of the order
of 100×10^6 m^3. The cumulative loss of water until September, 1972, as
mentioned by Handin and Nelson (1973), must be of the order of $10,000 \times
10^6$ m^3. Lake Powell is crossed by two monoclines and four minor folds of
Early Eocene age and by eight normal faults of Late Eocene age or younger.
According to Handin and Nelson, the water loss is due to outward and
downward percolation through open fissures raising the water pressure by
about 15 bars in these fractures and thereby reducing the normal effective
stress by the same amount. In spite of this definite invasion of pre-existing
faults by water, Lake Powell has shown no seismicity. A search of the
seismograms written by a set of three-component Benioff seismometers operating at the site since 1960 showed no event which could be assigned to the
lake. A high-gain array of seismometers which operated for four days on the
lake also did not reveal any event of magnitude zero or greater to have
occurred in or around the lake. Handin and Nelson (1973) tentatively concluded that the ambient effective stress difference in the Lake Powell region

was too small to have caused slip on the pre-existing slip and/or fault planes to cause tremors even with augmented pore pressures.

Detailed examination carried out on large artificial aseismic lakes impounded on faulted and fractured basins having competent rocks would probably reveal that the absence of sufficiently high ambient effective-stress differences is another major factor for their aseismic behavior.

Lomnitz (1974) has drawn attention to another factor which could explain the non-interference with the seismic status when the reservoirs are impounded in regions of moderate to high background seismicity. He cited the examples of two large dams, the Infiernillo and Malpaso Dams in western and southern Mexico, which have not evidenced any local associated seismicity, although sufficient seismic surveillance existed due to the operation of a short-period vertical seismometer. The ambient stress fluctuations in these areas are probably so high that the possible trigger effect of the reservoir becomes comparatively negligible; hence, can not interfere with the seismic status of the region. Although there are some examples of increase in small-magnitude seismic activity associated with large reservoirs in active regions, such as Japan, the effect in such regions remains small in terms of background seismicity. It has been generally observed that reservoirs located in highly seismic regions have shown no enhancement of local seismicity.

From the foregoing it appears that one or more of the following three factors may be responsible for the aseismic behavior of large artificial lakes:

(1) The presence of incompetent strata in the basement.

(2) The absence of large ambient stress differences in regions which are not critically stressed.

(3) High background seismicity rendering the possible trigger effect of the artificial lakes ineffective.

REFERENCES

Adams, R.D., 1969. Seismic effects at Mangla Dam. *UNESCO Rep.* No. 975, 12 pp.

Adams, R.D., 1974. The effect of Lake Benmore on local earthquakes. Paper presented at *Int. Colloq. on Seismic Effects of Reservoir Impounding*, The Royal Society, London, March, 1973. See also *Eng. Geol.*, 8: 155—169.

Adams, R.D. and Asghar Ahmed, 1969. Seismic effects at Mangla Dam. *Nature*, 222: 1153—1155.

Adams, R.D., Gough, D.I. and Muirhead, K.J., 1973. Seismic surveillance of artificial reservoirs. UNESCO Working Group on Seismic Phenomena Associated with Large Reservoirs, *Annexure 1, Rep. 3rd Meet.*, March, 1973, London, 8 pp.

Allen, C.R., Amand, P.S., Richter, C.F. and Nordquist, J.M., 1965. Relationship between seismicity and geologic structure in the southern California region. *Bull. Seismol. Soc. Am.*, 55: 753—797.

Anderson, R.E., 1971. Thin skin distension in Tertiary rocks of southeastern Nevada. *Bull. Geol. Soc. Am.*, 82: 43—58.

Anderson, R.E., 1973a. Large-magnitude Late Tertiary strike-slip faulting north of Lake Mead, Nevada. *U.S. Geol. Surv. Prof. Paper*, No. 794, 18 pp.

Anderson, R.E., 1973b. Late Cenozoic tectonic setting of Lake Mead, Nevada, Arizona, U.S.A. *Int. Colloq. on Seismic Effects of Reservoir Impounding*, March, 1973. The Royal Society, London, pp. 51—52 (summaries).

Anderson, R.E., Longwell, C.R., Armstrong, R.L. and Marvin, R.F., 1972. Significance of K—Ar ages of Tertiary rocks from the Lake Mead region, Nevada, Arizona. *Bull. Geol. Soc. Am.*, 83: 273—288.

Archer, C.B. and Allen, N.J., 1969. *A Catalogue of Earthquakes in the Lake Kariba Area, 1959—1968*. Meteorological Services, Salisbury, 35 pp.

Auden, J.B., 1972. Seismicity and reservoirs. *Comments Earth Sci., Geophys.*, 2: 149—150.

Balakrishna, S. and Gowd, T.N., 1970. Role of fluid pressure in the mechanics of transcurrent faulting at Koyna (India). *Tectonophysics*, 9: 301—321.

Banghar, A.R., 1972. Focal mechanisms of Indian earthquakes. *Bull. Seism. Soc. Am.*, 62: 603—608.

Bāth, M., 1965. Lateral inhomogeneities in the upper mantle. *Tectonophysics*, 2: 483—514.

Berg, E., 1968. Relation between earthquake foreshocks stress and main shocks. *Nature*, 219: 1141—1143.

Bernaix, J., 1967. *Etude géotechnique de la Roche de Malpasset*. Dunod, Paris, 215 pp.

Bhaskar Rao, V., Murty, B.V.S. and Murty, A.V.S.S., 1969. Some geological and geophysical aspects of the Koyna (India) earthquake, December 1967. *Tectonophysics*, 7: 265—271.

Bieniawski, Z.T., 1967. Mechanism of brittle fracture of rocks, 1 and 2. *Int. J. Rock Mech. Min. Sci.*, 4: 395—423.

Bond, G., 1953. The origin of thermal and mineral waters in the Middle Zambezi Valley and adjoining territory. *Trans Geol. Soc. S. Afr.*, 56: 131—148.

Bond, G., 1960. *The Geology of the Middle Zambezi Valley*. Unpublished manuscript.

Bowker, A.H. and Lieberman, G.J., 1959. *Engineering Statistics*. Prentice-Hall, Engle-
    wood Cliffs, N.J., 585 pp.
Brace, W.F. and Martin, R.J. 1968. A test of the law of effective stress for crystalline
    rocks of low porosity. *Int. J. Rock Mech. Min. Sci.*, 5: 415—426.
Brahmam, N.K. and Negi, J.G., 1973. Rift valleys beneath Deccan Traps (India). *Geo-
    phys. Res. Bull.* (India), 11: 207—237.
Brazee, R.J., 1969. Further reporting on the distribution of earthquakes with respect to
    magnitude (m_b). *Earthquake Notes*, 40: 49—51.
Bruhn, R.W., 1972. A study of the effects of pore pressure on the strength and deform-
    ability of Berea Sandstone in triaxial compression. Corps of Engineers, Missouri River
    Division Laboratory, *Tech. Rep. MRDL 1-T2*, 114 pp.
Bune, V.I. 1961. Some results of a detailed study of seismic conditions in the Stalinabad
    region in 1955—1959. *Bull. (Izv.) Acad. Sci., U.S.S.R., Geophys. Ser.*, 3: 237—242
    (English translation).
Byerlee, J.D., 1968. Brittle-ductile transition in rocks. *J. Geophys. Res.*, 73: 4741—4750.
Byerlee, J.D., 1971. Mechanical behaviour of Weber Sandstone. *EOS, Trans. Am. Geo-
    phys. Union*, 52: 343 (abstract).

Caloi, P., 1966. The results of geodynamic investigations in the Vajont's Gorge. *Ann.
    Geofis.* (Rome), 19: 1—74 (in Italian).
Caloi, P., 1970. How nature reacts on human intervention — responsibilities of those who
    cause and who interpret such reaction. *Ann. Geofis.* (Rome), 23: 283—305.
Carder, D.S., 1945. Seismic investigations in the Boulder Dam area, 1940—1944, and the
    influence of reservoir loading on earthquake activity. *Bull. Seismol. Soc. Am.*, 35:
    175—192.
Carder, D.S., 1968. Reservoir and local earthquakes. In: *Engineering Geology and Soils
    Engineering*. Proc. 6th Annu. Symp., Boise, Idaho, 1968, Dep. of Highways, Boise,
    Idaho, pp. 225—241.
Carder, D.S., 1970. Reservoir loading and local earthquakes in engineering seismology —
    the works of man. In: W.M. Adams (Editor), *Engineering Geology Case Histories,
    No. 8*. Geological Society of America, Denver, Colo., pp. 51—61.
Chandra, U., 1970. Table for the angle of incidence at the focus for S waves based on
    Randall's revised S-tables. *Earthquake Notes*, 41: 35—43.
Chaterji, G.C., 1969. Mineral and thermal waters of India. *Proc. Int. Geol. Congr., Rep.
    23rd Sess.*, Prague, 1968, 19: 21—43.
Chaudhury, H.M. and Srivastava, H.N., 1973. The time of occurrence and the magnitude
    of the largest aftershock over India. *Pure Appl. Geophys.* 105: 770—780.
Chung-kang, S., Hou-chun, C., Chu-han, C., Li-sheng, H., Tzu-chiang, L., Chen-yung, Y.,
    Ta-chun, W. and Hsueh-hai, L., 1974. Earthquakes induced by reservoir impounding
    and their effect on the Hsinfengkiang Dam. *Sci. Sinica*, 17 (2): 239—272.
Committee of Experts, 1968. *Report on the Koyna Earthquake of December 11, 1967 —
    1 and 2*. Gov. of India Press, New Delhi, 75 pp.
Comninakis, P., Drakopoulos, J., Moumoulidis, G. and Papazachos, B.C., 1968. Fore-
    shock sequences of the Kremasta earthquake and their relation to the water loading of
    the Kremasta artificial lake. *Ann. Geofis.* (Rome), 21: 39—71.
Cornet, C. and Fairhurst, C., 1972. Variation of pore volume in disintegrating rock. *Proc.
    Symp. on Percolation Through Fissured Rocks*. Deutsche Gesellschaft für Erd- und
    Grundbau, Stuttgart, T2-A1: 1—8.
Coulomb, C.A., 1776. Essai sur une application des règles de maximis et minimis à
    quelques problèmes de statique, relatifs à l'architecture. *Acad. R. Sci., Paris, Mém.
    Math. Phys.*, V 7 (1779): 343—382.

Das, S.R. and Ray, A.K., 1972. A note on the photogeological study of Koyna region and part of the west coast in Satara, Ratnagiri and Kolaba districts, Maharashtra. *Indian Miner.*, Geol. Surv. of India, 26: 76—79.

De la Cruz, R.V. and Raleigh, C.B., 1972. Absolute stress measurements at the Rangely anticline, northwestern Colorado. *Int. J. Rock Mech. Min. Sci.*, 9: 625—634.

De Sitter, L.U., 1965. *Structural Geology. Series in the Geological Sciences.* McGraw-Hill, New York, N.Y., 552 pp.

Dewey, J.F. and Bird, J.M., 1970. Mountain belts and the new global tectonics. *J. Geophys. Res.*, 75: 2625—2646.

Dieterich, J.H., Raleigh, C.B. and Bredehoeft, J.D., 1972. Earthquake triggering by fluid injection at Rangely — Colorado. *Proc. Symp. on Percolation through Fissured Rocks.* Deutsche Gesellschaft für Erd- und Grundbau, Stuttgart, T2-B: 1—12.

Drakopoulos, J., 1973. Seismic activity close to Kremasta—Kastraki Dams, Greece, and related problems. *Int. Colloq. on Seismic Effects of Reservoir Impounding*, March, 1973. The Royal Society, London, pp. 9—10 (summaries).

Drysdall, A.R. and Weller, R.K., 1966. Karoo sedimentation in northern Rhodesia. *Trans. Geol. Soc. S. Afr.*, 69: 39—69.

Dutta, T.K., 1969. A note on the source parameters of the Koynanagar earthquake of December 10, 1967. *Bull. Seismol. Soc. Am.*, 59: 935—944.

Evans, M.D., 1966. Man made earthquakes in Denver. *Geotimes*, 10: 11—17.

Evernden, J.F., 1970. Study of regional seismicity and associated problems. *Bull. Seismol. Soc. Am.*, 60: 393—446.

Fairhurst, C., 1973a. Mechanics of stable and unstable rock fracture, and influence of pore fluid pressure. *Int. Colloq. on Seismic Effects of Reservoir Impounding*, March, 1973. The Royal Society, London, p. 38 (summaries).

Fairhurst, C., 1973b. Determination of rock stresses at depth. *Int. Colloq. on Seismic Effects of Reservoir Impounding*, March, 1973. The Royal Society, London, p. 70 (summaries).

Fairhurst, C. and Roegiers, J.C., 1972. Estimation of rock mass permeability by hydraulic fracturing — a suggestion. *Proc. Symp. on Percolation through Fissured Rocks.* Deutsche Gesellschaft für Erd- und Grundbau, Stuttgart, D2: 1—5.

Fontsere, E., 1963. The earthquakes at Catalanes from 1955 to 1962. *Reas Acad. Ci. Artes Barcelona, Observ., Fabra*, 9 pp. (in Spanish).

Friedman, M., 1964. Petrofabric techniques for the determination of principal stress in the earth's crust. In: W.R. Judd (Editor), *State of Stress in the Earth's Crust.* American Elsevier, New York, N.Y., pp. 451—552.

Galanopoulos, A.G., 1965. Evidence for the seat of the strain producing forces. *Ann. Geofis.* (Rome), 18: 399—409.

Galanopoulos, A.G., 1967a. The large conjugate fault system and the associated earthquake activity in Greece. *Ann. Geol. Pays Helleniques* (Athens), 18: 119—134.

Galanopoulos, A.G., 1967b. The influence of the fluctuation of Marathon Lake elevation of local earthquake activity in the Attica Basin area. *Ann. Geol. Pays Helleniques* (Athens), 18: 281—306.

Geological Survey of India, 1968. *A Geological Report on the Koyna Earthquake of 11th December 1967.* Calcutta, 242 pp.

Gibbs, J.F., Healy, J.H., Raleigh, C.B. and Coakley, J., 1973. Seismicity in the Rangely, Colorado, area: 1962—1970. *Bull. Seismol. Soc. Am.*, 63: 1557—1570.

Goguel, J., 1973. Implications tectoniques des séismes provoqués par le remplissage de réservoirs. *Int. Colloq. on Seismic Effects of Reservoir Impounding*, March, 1973. The Royal Society, London, pp. 62—63 (summaries).

Goodman, R.E., 1973. Pore pressure in intact and jointed rocks induced by deformation. *Int. Colloq. on Seismic Effects of Reservoir Impounding*, March, 1973. The Royal Society, London, p. 41 (summaries).

Gorbunova, I.V., Kondorskaya, N.V. and Landyreva, N.S., 1970. On the determination of the extent of an Indian shock origin by kinematic data. *Geophys. J.*, 20: 457—471.

Gorshkov, G.P., 1963. The seismicity of Africa. In: *A Review of the National Resources of the African Continent*. UNESCO, Int. Document Service, Columbia Univ. Press, New York, N.Y., pp. 101—151.

Gough, D.I., 1969. Incremental stress under a two-dimensional artificial lake. *Can. J. Earth Sci.*, 6: 1067—1075.

Gough, D.I. and Gough, W.I., 1970a. Stress and deflection in the lithosphere near Lake Kariba, 1. *Geophys. J.*, 21: 65—78.

Gough, D.I. and Gough, W.I., 1970b. Load-induced earthquakes at Lake Kariba, 2. *Geophys. J.*, 21: 79—101.

Gough, D.I. and Gough, W.I., 1973. Stress under Cabora Bassa. *Int. Colloq. on Seismic Effects of Reservoir Impounding*, March, 1973. The Royal Society, London, p. 60 (summaries).

Gourinard, Y., 1952. La géologie et les problèmes de l'eau en Algérie, 1. Le barrage de l'Oued Fodda. *19th Congr. Geol. Int.*, C. R., 1: 155—181.

Green, R.W.E., 1974. Seismic activity observed at the Hendrik Verwoerd Dam. Paper presented at *Int. Colloq. on Seismic Effects of Reservoir Impounding*, The Royal Society, London, March, 1973. Communicated to *J. Eng. Geol.*

Gubin, I.E., 1969. Koyna earthquake of 1967. *Bull. Inst. Seism. Earthquake Eng.*, 6: 45—62.

Guha, S.K., Gosavi, P.D., Varma, M.M., Agarwal, S.P., Padale, J.G. and Marwadi, S.C., 1968. Recent seismic disturbances in the Koyna Hydroelectric Project, Maharashtra, India, 1. *Rep. C.W.P.R.S.*, 16 pp.

Guha, S.K., Gosavi, P.D., Varma, M.M., Agarwal, S.P., Padale, J.G. and Marwadi, S.C., 1970. Recent seismic disturbances in the Shivajisagar Lake area of the Koyna Hydroelectric Project. Maharashtra, India, 2. *Rep. C.W.P.R.S.*, 25 pp.

Guha, S.K., Gosavi, P.D., Agarwal, B.N.P., Padale, J.G. and Marwadi, S.C., 1974. Case histories of some artificial crustal disturbances. Paper presented at *Int. Colloq. on Seismic Effects of Reservoir Impounding*, The Royal Society, London, March, 1973. See also *Eng. Geol.*, 8: 59—77.

Gupta, H.K. and Rastogi, B.K., 1974a. Will another damaging earthquake occur in Koyna? *Nature*, 248: 215—216.

Gupta, H.K. and Rastogi, B.K., 1974b. Investigations of the behaviour of reservoir associated earthquakes. Paper presented at *Int. Colloq. on Seismic Effects of Reservoir Impounding*, The Royal Society, London, March, 1973. See also *Eng. Geol.*, 8: 29—38.

Gupta, H.K., Narain, H., Rastogi, B.K. and Mohan, I., 1969. A study of the Koyna earthquake of December 10, 1967. *Bull. Seismol. Soc. Am.*, 59: 1149—1162.

Gupta, H.K., Mohan, I. and Narain, H., 1970. The Godavari Valley earthquake sequence of April 1969. *Bull. Seismol. Soc. Am.*, 60: 601—615.

Gupta, H.K., Rastogi, B.K. and Narain, H., 1971. The Koyna earthquake of December 10: a multiple seismic event. *Bull. Seismol. Soc. Am.*, 61: 167—176.

Gupta, H.K., Rastogi, B.K. and Narain, H., 1972a. Common features of the reservoir-associated seismic activities. *Bull. Seismol. Soc. Am.*, 62: 481—492.

Gupta, H.K., Rastogi, B.K. and Narain, H., 1972b. Some discriminatory characteristics of earthquakes near the Kariba, Kremasta and Koyna artificial lakes. *Bull. Seismol. Soc. Am.*, 62: 493—507.

Gupta, H.K., Mohan, I. and Narain, H., 1972c. The Broach earthquake of March 23, 1970. *Bull. Seismol. Soc. Am.*, 62: 47—61.

Gupta, H.K., Rastogi, B.K. and Narain, H., 1973. A study of earthquakes in the Koyna region and common features of the reservoir-associated seismicity. In: W.C. Ackermann, G.F. White and E.B. Worthington (Editors), *Geophysical Monograph Series No. 17*. American Geophysical Union, Washington, D.C., pp. 455—467.

Gupta, M.L. and Sukhija, B.S., 1974. Preliminary studies of some geothermal areas in India. *Geothermics*, 3: 105—112.

Gutenberg, B. and Richter, C.F., 1954. *Seismicity of the Earth*. Princeton Univ. Press, Princeton, N.J., 2nd ed., 273 pp.

Gutenberg, B. and Richter, C.F., 1956. Magnitude and energy of earthquakes. *Ann. Geofis.* (Rome), 9: 1—15.

Hadsell, F.A., 1968. History of earthquake activity in Colorado. *Q. Colo. Sch. Mines*, 63: 57—72.

Hagiwara, T. and Ohtake, M., 1972. Seismic activity associated with the filling of the reservoir behind the Kurobe Dam, Japan, 1963—1970. *Tectonophysics*, 15: 241—254.

Haimson, B.C., 1972. Earthquake related stresses at Rangely, Colorado. *Proc. 14th Symp. on Rock Mechanics*, State College, Pennsylvania State Univ., Pasadena, Pa.

Haimson, B. and Fairhurst, C., 1970. In-situ stress determination at great depth by means of hydraulic fracturing. In: *Rock Mechanics — Theory and Practice*. American Institute of Mining, Metallurgy and Petroleum Engineers, New York, pp. 559—584.

Handin, J., 1958. Effects of pore pressure on the experimental deformation of sedimentary rocks under high pressure: tests at room temperature on dry samples. *Am. Assoc. Pet. Geol. Bull.*, 41: 1—50.

Handin, J. and Nelson, R.A., 1973. Why is Lake Powell aseismic? *Int. Colloq. on Seismic Effects of Reservoir Impounding*, March, 1973. The Royal Society, London, p. 53—55 (summaries).

Handin, J. and Raleigh, C.B., 1972. Man-made earthquakes and earthquake control. *Proc. Symp. on Percolation through Fissured Rocks*. Deutsche Gesellschaft für Erd- und Grundbau, Stuttgart, T2D: 1—10.

Healy, J.H., Rubey, W.W., Griggs, D.T. and Raleigh, C.B., 1968. The Denver earthquakes. *Science*, 161: 1301—1310.

Hitchon, B., 1958. The geology of the Kariba area. *North Rhod. Geol. Surv., Rep. No. 3*, 41 pp.

Hoffmann, R.B., 1969. Earthquake predictions from fault movement and strain precursors in California. In: L. Mansinha, D.E. Smylie and A.E. Beck (Editors), *Earthquake Displacement Fields and the Rotation of the Earth*. D. Reidel, Dordrecht, pp. 234—245.

Honda, H., 1957. The mechanism of the earthquake. *Sci. Rep., Tohoku Univ., Ser. 5 (Geophysics)*, 9: 1—46 (supplement).

Howells, D.A., 1973. The time for a significant change of pore pressure. Paper presented at *Int. Colloq. on Seismic Effects of Reservoir Impounding*, The Royal Society, London, March, 1973. See also *Eng. Geol.*, 8: 135—138.

Hubbert, M.K. and Rubey, W.W., 1959. Role of fluid pressure in mechanics of overthrust faulting, 1. *Bull. Geol. Soc. Am.*, 70: 115—166.

Hubbert, M.K. and Willis, D.G., 1957. Mechanics of hydraulic fracturing. *Trans. Am. Inst. Min. Metall. Eng.*, 210: 153—166.

Isacks, B. and Oliver, J., 1964. Seismic waves with frequencies from 1 to 100 cycles per second recorded in a deep mine in northern New Jersey. *Bull. Seismol. Soc. Am.*, 54: 1941—1979.

Ishimoto, M. and Iida, K., 1939. Observations sur les séismes enregistrés par le microsismographe construit dernièrement. *Bull. Earthquake Res. Inst.*, 17: 443—478.

Jaeger, J.C., 1956. *Elasticity, Fracture and Flow*. Methuen, London, 152 pp.

Jai Krishna, Chandra Sekharan, A.R. and Saini, S.S., 1969. Analysis of Koyna accelerograms of December 11, 1967. *Bull. Seismol. Soc. Am.*, 59: 1719—1731.

Jai Krishna, Arya, A.S. and Kumar, K., 1970. Importance of isoforce lines of an earthquake with special reference to Koyna earthquake of December 11, 1967. *4th Symp. on Earthquake Engineering*, Roorkee Univ., Roorkee, November, 1970. Sarita Prakashan, Meerut, pp. 1—13.

Jones, A.E., 1944. Earthquake magnitudes, efficiency of stations and perceptibility of local earthquakes in the Lake Mead area. *Bull. Seismol. Soc. Am.*, 34: 161—173.

Jouanna, P., 1972. Laboratory tests on the permeability of mica schist samples under applied stresses. *Proc. Symp. on Percolation through Fissured Rocks*. Deutsche Gesellschaft für Erd- und Grundbau, Stuttgart, T2-F1: 1—11.

Kailasam, L.N. and Murthy, B.G.K., 1969. Geophysical investigations in earthquake affected areas of Koyna, Satara District, Maharashtra. *Mem. Geol. Surv. India*, 100: 117—122.

Kailasam, L.N., Pant, P.R. and Lahiri, S.M., 1969. Seismic investigations in the earthquake affected areas of Koyna and neighbourhood, Satara district, Maharashtra. *Mem. Geol. Surv. India*, 100: 123—126.

Kailasam, L.N., Murthy, B.G.K. and Chayanulu, A.Y.S.R., 1972. Regional gravity studies of the Deccan Trap areas of Peninsular India. *Current Science*, 41 (11): 403—407.

Karnik, V., 1969. *Seismicity of the European Area*. D. Reidel, Dordrecht, 364 pp.

Khattri, K.N., 1970. The Koyna earthquake — seismic studies. *4th Symp. on Earthquake Engineering*, Roorkee Univ., Roorkee, November, 1970. Sarita Prakashan, Meerut, pp. 369—374.

King, L.C., 1962. *Morphology of the Earth*. Oliver and Boyd, Edinburgh, 599 pp.

Kirkpatrick, I.M. and Robertson, I.O.M., 1968. A geologic reconnaissance of the Makuti—Kariba road. *Rhod. Geol. Surv., Rep.*, 16 pp.

Knill, J.L. and Jones, K.S., 1965. The recording and interpretation of geological conditions in the foundations of the Rosieres, Kariba and Latiyan Dams. *Geotechnique*, 158: 94—124.

Krishnan, M.S., 1960. *Geology of India and Burma*. Higginbothms, Madras, 604 pp.

Krishnan, M.S., 1966. Tectonics of India. *Bull. Indian Geophys. Union*, 3: 1—36.

Krsmanovic, D., 1967. Initial and residual shear strength of hard rocks. *Geotechnique*, 17: 145—160.

Lara, J.M. and Sanders, J.I., 1970. *The 1963—64 Lake Mead Survey*. Bureau of Reclamation, Denver, Colo., Rep. REC-OCE-70-21. 174 pp.

Lee, W.H.K. and Cox, C.S., 1966. Time variation of ocean temperatures and its relation to internal waves and oceanic heat flow measurements. *J. Geophys. Res.*, 71: 2101—2111.

Lee, W.H.K. and Raleigh, C.B., 1969. Fault-plane solution of the Koyna (India) earthquake. *Nature*, 223: 172—173.

Le Tirant, P. and Baron, G., 1972. Flow in fissured rocks and effective stresses — application to hydrocarbon production and to hydraulic fracturing. *Proc. Symp. on Percolation through Fissured Rocks*. Deutsche Gesellschaft für Erd- und Grundbau, Stuttgart, T2-K: 1—24.

Lombardi, J., 1967. Quelques problèmes de mécanique des roches étudiés lors de la construction du barrage de Contra (Cerzasca). *9th Congr. on Large Dams*, Istanbul, 1967, Paper R-15. Commission Internationale des Grands Barrages, Paris, pp. 235—252.

Lomnitz, C., 1974. Earthquake and reservoir impounding: state of the art. Paper presented at *Int. Colloq. on Seismic Effects of Reservoir Impounding*, The Royal Society, London, March, 1973. See also *Eng. Geol.*, 8: 191—198.

Longwell, C.R., 1936. Geology of the Boulder reservoir floor. Arizona, Nevada. *Bull. Geol. Soc. Am.*, 47: 1393—1476.

Longwell, C.R., 1963. Reconnaissance geology between Lake Mead and Arvis Dam, Arizona, Nevada. *U.S. Geol. Surv. Prof. Paper*, No. 374-E, 51 pp.

Longwell, C.R., Pampeyan, E.H., Bowyer, B. and Roberts, R.J., 1965. Geology and mineral deposits of Clark country, Nevada. *Nevada Bur. Min. Bull.*, 62: 218.

Mane, P.M., 1967. Earth tremors in the Koyna Project area. *9th Congr. on Large Dams, Istanbul, 1967. Commission Internationale des Grands Barrages, Paris*, pp. 509—518.

Mann, A.G., 1967. Investigations on jointing associated with the Deka fault, Rhodesia. Unpublished Rep., pp. 24—28.

McEvilly, T.V. and Casaday, K.B., 1967. The earthquake sequence of September 1965 near Antioch, California. *Bull. Seismol. Soc. Am.*, 57: 113—124.

McEvilly, T.V., Bakun, W.W.H. and Casaday, K.B., 1967. The Parkfield, California, earthquakes of 1966. *Bull. Seismol. Soc. Am.*, 57: 1221—1224.

McGinnis, L.D., 1963. Earthquakes and crustal movements as related to water load in the Mississippi valley region. *Ill. State Geol. Surv., Circ.* No. 344, 20 pp.

Mickey, W.V., 1973a. Reservoir seismic effects. In: W.C. Ackermann, G.F. White and E.B. Worthington (Editors), *Geophysical Monograph Series No. 17*. American Geophysical Union, Washington, D.C., pp. 472—479.

Mickey, W.V., 1973b. Seismic effects of reservoir impounding in the United States. *Int. Colloq. on Seismic Effects of Reservoir Impounding*, March, 1973. The Royal Society, London, pp. 19—20 (summaries).

Mindling, A., 1971. A summary of data relating to land subsidence in Las Vegas Valley. *Desert Res. Inst. Rep.*, Nevada Univ., Reno, Nev. (unpublished report to the U.S. Atomic Energy Commission).

Miyamura, S., 1962. Magnitude—frequency relation of earthquakes and its bearing on geotectonics. *Proc. Japan Acad.*, 38: 27—30.

Mizoue, M., 1967. Variation of earthquake energy release with depth, 1. *Bull. Earthquake Res. Inst.*, 45: 679—709.

Mogi, K., 1962a. On the time distribution of aftershocks accompanying the recent major earthquakes in and near Japan. *Bull. Earthquake Res. Inst.*, 40: 107—124.

Mogi, K., 1962b. Study of elastic shocks caused by the fracture of heterogeneous materials and its relations to earthquake phenomena. *Bull. Earthquake Res. Inst.*, 40: 125—173.

Mogi, K., 1963a. The fracture of a semi-infinite body caused by an inner stress origin and its relation to the earthquake phenomena (second paper). The case of the materials having some heterogeneous structures. *Bull. Earthquake Res. Inst.*, 41: 595—614.

Mogi, K., 1963b. Some discussions on aftershocks, foreshocks and earthquake swarms — the fracture of a semi-infinite body caused by an inner stress origin and its relation to the earthquake phenomena (third paper). *Bull. Earthquake Res. Inst.*, 41: 615—658.

Mogi, K., 1967a. Earthquakes and fractures. *Tectonophysics*, 5: 35—55.

Mogi, K., 1967b. Regional variations in magnitude—frequency relation of earthquakes. *Bull. Earthquake Res. Inst.*, 45: 313—325.

Mohr, O.C., 1882. Ueber die Darstellung des Spannungszustandes und des Deformationszustandes eines Korperelementes und über die Anwendung derselben in der Festigkeitslehre. *Der Civilingenieur*, 5 (28): 113—156.

Morgenstern, N.R. and Guther, H., 1972. Seepage into an excavation in a medium possessing stress dependent permeability. *Proc. Symp. on Percolation through Fissured Rocks*. Deutsche Gesellschaft für Erd- und Grundbau, Stuttgart, T2-C1: 1—15.

Muirhead, K.J., Cleary, J.R., Simpson, D.W., 1973. Seismic activity associated with the filling of Talbingo Reservoir. *Int. Colloq. on Seismic Effects of Reservoir Impounding*, March, 1973. The Royal Society, London, pp. 17 (summaries).

Nakano, H., 1923. Notes on the nature of forces which give rise to earthquake motions. *Seismol. Bull., Centre Meteorol. Observ. Japan.*, 1: 91—120.

Narain, H. and Gupta, H.K., 1968a. Observations on Koyna earthquake. *J. Indian Geophys. Union*, 5: 30—34.

Narain, H. and Gupta, H.K., 1968b. The Koyna earthquake, *Nature*, 217: 1138—1139.

Narain, H. and Gupta, H.K., 1968c. How and why of Koyna. *Sci. To-day*, February: 47—51.

Nason, R.D. and Tocher, D., 1969. Measurement of movement on the San Andreas Fault. In: L. Mansinha, D.E. Smylie and A.E. Beck (Editors), *Earthquake Displacement Fields and the Rotation of the Earth*. D. Reidel, Dordrecht, pp. 246—254.

National Academy of Sciences/National Academy of Engineering, U.S.A., 1972. *Report: Earthquakes Related to Reservoir Filling*. Division of Earth Sciences, National Research Council, Washington, D.C., 24 pp.

Noorishad, J., Witherspoon, P.A. and Maini, Y.M.T., 1972. The influence of fluid injection on the state of stress in the earth's crust. *Proc. Symp. on Percolation through Fissured Rocks*. Deutsche Gesellschaft für Erd- und Grundbau, Stuttgart, T2-H: 1—11.

Nur, A. and Booker, J.R., 1972. Aftershocks caused by pore fluid flow? *Science*, 175: 885—887.

Nuttli, O.W., 1969. Table of angles of incidence of P-waves at focus, calculated from 1968 P-tables. *Earthquake Notes*, 40: 21—25.

Pakiser, L.C., Eaton, J.P., Healy, J.H. and Raleigh, C.B., 1969. Earthquake prediction and control. *Science*, 166: 1467—1474.

Papazachos, B.C., 1971. Aftershock activity and aftershock risk in the area of Greece. *Ann. Geofis.* (Rome), 24: 439—456.

Papazachos, B.C., 1973. The time distribution of the reservoir-associated foreshocks and its importance to the prediction of the principal shock. *Bull. Seismol. Soc. Am.*, 63: 1973—1978.

Papazachos, B.C., 1974. On the relation between certain artificial lakes and the associated seismic sequences. Paper presented at *Int. Colloq. on Seismic Effects of Reservoir Impounding*, The Royal Society, London, March, 1973. See also *Eng. Geol.*, 8: 39—48.

Papazachos, B., Delibasis, N., Liapis, N., Moumoulidis, G. and Purcasu, G., 1967. Aftershock sequences of some large earthquakes in the region Greece. *Ann. Geofis.* (Rome), 20: 1—93.

Papazachos, B., Comninakis, P., Drakopoulos, J. and Moumoulides, G., 1968. *Foreshock and Aftershock Sequences of the Cremasta Earthquake and their Relation to the Water Loading of the Cremasta Artificial Lake*. Seismological Institute, Natl. Observ. of Athens, Athens, 27 pp.

Raleigh, C.B., 1972. Underground waste management and environmental inplications. *Am. Assoc. Pet. Geol., Mem.*, 18: 273—279.

Raleigh, C.B., Healy, J.H. and Bredehoeft, H.D., 1972. Faulting and crustal stress at Rangely, Colorado. In: H.C. Heard, I.Y. Borg, N.L. Carter and C.B. Raleigh (Editors), *Geophysical Monograph No. 16*. American Geophysical Union, Washington, D.C., pp. 275—284.

Raphael, J.M., 1954. Crustal disturbance in the Lake Mead area. In: *Engineering Monographs, No. 14*. U.S. Bureau of Reclamation, Denver, Colo., p. 14.

Richter, C.F., 1958. *Elementary Seismology*. W.H. Freeman and Co., San Francisco, Calif., and London, 768 pp.

Riznichenko, V.Y., 1959. On quantitative determination and mapping of seismic activity. *Ann. Geofis.* (Rome), 12: 227—237.

Rodatz, W. and Wittke, W., 1972. interaction between deformation and percolation in fissured anisotropic rock. *Proc. Symp. on Percolation through Fissured Rocks.* Deutsche Gesellschaft für Erd- und Grundbau, Stuttgart, T2-I, 1—18.

Rogers, A.M. and Gallanthine, S.K., 1974. Seismic study of earthquakes in the Lake Mead region. *Rep., Environmental Research Corporation, U.S.A.,* Contract No. 14-08-0001-13069.

Roksandic, M.M., 1970. Influence de la charge d'un reservoir sur l'activité séismique. *Proc. 2nd Congr. Int. Soc. Rock Mechanics,* 3: 8—12.

Rothé, J.P., 1968. Fill a lake, start an earthquake. *New Scientist,* 39: 75—78.

Rothé, J.P., 1969. Earthquake reservoir loadings. *4th World Conf. on Earthquake Engineering,* Santiago, 1969, Preprints, A-1: 28—38 and addendum.

Rothé, J.P., 1970. Séismic artificiels (man-made earthquakes). *Tectonophysics,* 9: 215—238.

Rothé, J.P., 1972. The seismicity of France from 1961 to 1970. *Ann. Inst. Physique du Globe,* 9: 3—134.

Rothé, J.P., 1973. Man-made lakes: their problems and environmental effects — a geophysics report. In: W.C. Ackermann, G.F. White and E.B. Worthington (Editors), *Geophysical Monograph No. 17.* American Geophysical Union, Washington, D.C., pp. 441—454.

Rummel, F. and Gowd, T.N., 1973. Effect of interstitial fluid pressure on faulting in Ruhr-Sandstone specimens. *Int. Colloq. on Seismic Effects of Reservoir Impounding,* March, 1973. The Royal Society, London, pp. 39 (summaries).

Sato, Y. and Skoko, D., 1965. Optimum distribution of seismic observation points, 2. *Bull. Earthquake Res. Inst.,* 43: 451—457.

Scholz, C.H., 1968. The frequency—magnitude relation of microfracturing in rock and its relation to earthquakes. *Bull. Seismol. Soc. Am.,* 58: 399—415.

Scholz, C.H., Sykes, L.R. and Aggarwal, Y.P., 1973. Earthquake prediction: a physical basis. *Science,* 181: 803—810.

School of Research and Training in Earthquake Engineering, 1968. *Behaviour of Structures in the Koyna Earthquake of December 11, 1967.* Unpublished report, Univ. of Roorkee, Roorkee.

Simon, R.B., 1969. Seismicity of Colorado: consistency of recent earthquakes with those of historical record. *Science,* 165: 897—899.

Simon, R.B., 1972. *Seismicity, Geologic Atlas of the Rocky Mountain Region.* Rocky Mountain Association of Geologists, Denver, Colo., pp. 48—51.

Singh, D.D., Rastogi, B.K. and Gupta, H.K., 1975. Surface-wave radiation pattern and source parameters of the Koyna earthquake of December 10, 1967. *Bull. Seismol. Soc. Am.,* 65: 711—731.

Sleigh, R.W., Worrall, C.C. and Shaw, G.H.L., 1969. Crustal deformation resulting from the imposition of a large mass of water. *Bull. Geodes.,* 93: 245—254.

Snow, D.T., 1968a. Hydraulic characteristics of fractured metamorphic rocks of Front and Rangely and implications to the Rocky Mountain Arsenal Well. *Q. Colo. Sch. Mines,* 63: 167—200.

Snow, D.T., 1968b. Fracture deformation and change of permeability and storage upon changes of fluid pressure. *Q. Colo. Sch. Mines,* 63: 201—244.

Snow, D.T., 1972. Geodynamics of seismic reservoirs. *Proc. Symp. on Percolation through Fissured Rocks.* Deutsche Gesellschaft für Erd- und Grundbau, Stuttgart, T2-J: 1—19.

Snow, D.T., 1974. The geologic, hydrologic and geomorphic setting of earthquakes at Lake Kariba. Paper presented at *Int. Colloq. on Seismic Effects of Reservoir Impounding,* The Royal Society, London, March, 1973.

Susstrunk, A., 1968. Erdstosse im Verzascatal beim Aufstau des Speeicherbekens, Vagorno. *Verh. Schweiz. Naturforsch. Ges.*, 89—92.

Sykes, L.R., 1967. Mechanism of earthquakes and nature of faulting on the mid-oceanic ridges. *J. Geophys. Res.*, 72: 2131—2153.

Sykes, L.R., 1970. Seismicity of the Indian Ocean and a possible nascent island arc between Ceylon and Australia. *J. Geophys. Res.*, 75: 5041—5055.

Sykes, L.R., Fletcher, J.P. and Sbar, M.L., 1973. Contemporary stresses intraplate earthquakes, and seismic risk associated with high-pressure fluid injection wells in New York State. *Int. Colloq. on Seismic Effects of Reservoir Impounding*, March, 1973. The Royal Society, London, pp. 46 (summaries).

Tandon, A.N., 1954. Study of the great Assam earthquake of August, 1950, and its aftershocks. *Ind. J. Meteorol. Geophys.*, 5: 95—137.

Tandon, A.N. and Chaudhury, H.M., 1968. Koyna earthquake of December, 1967. *India Meteorol. Dep., Sci. Rep.* No. 59, 12 pp.

Terra-Consult, Inc., 1965. *Engineering Geological Report on General Geology, Engineering Geology and Hydrogeology in the Kremasta Reservoir Area.* Hanover, 130 pp. (unpublished).

Thirlaway, H.I.S., 1963. Earthquake or explosion. *New Scientist*, 18: 311—315.

Timmel, K.E. and Simpson, D.W., 1973. Seismic events during filling of Talbingo Reservoir. Unpublished Rep., Australian National University, Canberra, A.C.T., pp. 27—33.

Timoshenko, S. and Goodier, J.N., 1951. *Theory of Elasticity*. McGraw-Hill, London and New York, N.Y., 2nd ed., 506 pp.

Tomita, H. and Utsu, T., 1968. Magnitude distribution of earthquakes in various regions of the world. *J. Fac. Sci., Hokkaido Univ.*, 19: 57—64 (in Japanese).

Trifunac, M.D. and Brune, J.N., 1970. Complexity of energy release during the Imperial Valley, California, earthquake of 1940. *Bull. Seismol. Soc. Am.*, 60: 137—160.

Tsai, Y-Ben and Aki, K., 1971. The Koyna, India, earthquake of December 10, 1967 (abstract). *Trans. Am. Geophys. Union*, 52: 277.

UNESCO Working Group on Seismic Phenomena Associated with Large Reservoirs, 1970. Report of the First Meeting, UNESCO Headquarters, Paris, December, 1970. *UNESCO Rep.* No. SC/CONF. 200/4, 7 pp.

UNESCO Working Group on Seismic Phenomena Associated with Large Reservoirs, 1973. Report of the Third Meeting, London, March, 1973. *UNESCO Rep.* No. SC-73/CONF. 625/1, 19 pp.

Utsu, T., 1961. A statistical study on the occurrence of aftershocks. *Geophys. Mag.* (Tokyo), 30 (4): 523—605.

Utsu, T., 1965. A method for determining the value of b in the formula $\log n = a - bM$, showing the magnitude—frequency relation for earthquakes. *Geophys. Bull., Hokkaido Univ.*, 13: 99—103.

Utsu, T., 1966. A statistical significance test of the difference in b-value between two earthquake groups. *J. Phys. Earth*, 14: 37—40.

Utsu, T., 1969. Aftershocks and earthquake statistics, I. Some parameters which characterize an aftershock sequence and their interrelations. *J. Fac. Sci., Hokkaido Univ., Ser. 7 (Geophysics)*, 3: 129—195.

Wadia, D.N., 1968. The Koyna earthquake — December, 1967. *J. Indian Geophys. Union*, 5: 6—8.

Walters, R.C.S., 1971. *Dam Geology*. Butterworth, London, 2nd ed., 470 pp.

Westergaard, H.M. and Adkins, A.W., 1934. *Deformation of Earth's Surface due to Weight of Boulder Reservoir*. U.S. Bureau of Reclamation, Denver, Colo., Tech. Mem. No. 422.

Wittke, W., 1973. Consideration for mechanical and hydraulic properties of jointed rock masses with numerical computations. *Int. Colloq. on Seismic Effects of Reservoir Impounding*, March, 1973. The Royal Society, London, pp. 50 (summaries).

Wyss, M. and Brune, J.N., 1967. The Alaska earthquake of 28, March 1964: a complex multiple rupture. *Bull. Seismol. Soc. Am.*, 57: 1017—1023.

Zienkiewiez, O.C. and Cheung, Y.K., 1967. *The Finite-Element Method in Structural and Continuum Mechanics*. McGraw Hill, New York, N.Y., 272 pp.

AUTHOR INDEX

SUBJECT INDEX